GOODWILL'S

100 AMAZING MAKE-IT-YOURSELF SCIENCE FAIR PROJECTS

Glen Vecchione

GOODWILL PUBLISHING HOUSE®
B-3 RATTAN JYOTI, 18 RAJENDRA PLACE
NEW DELHI -110008 (INDIA)

This special low priced Indian reprint is published by arrangement with **Sterling Publishing Company, Inc. New York, U.S.A.**

Published in India by
GOODWILL PUBLISHING HOUSE®
B-3 Rattan Jyoti, 18 Rajendra Place
New Delhi-110008 (INDIA)
Tel. : 25750801, 25820556
Fax : 91-11-25764396
E-mail : goodwillpub@vsnl.net
website : www.goodwillpublishinghouse.com

Printed at : Kumar Offset Printers, Delhi-110092

Contents

Introduction

We all love to see how things work. As long ago as 1652, Otto von Guericke, mayor of Magdeburg and dabbler in science, proved that even two teams of horses could not pull apart his vacuum-sealed ball. The German emperor was impressed. In 1783, the Montgolfier brothers launched their famous hot-air balloon from a Paris park in front of an audience of thousands. The 19th century inventors Alexander Graham Bell and Thomas Edison often demonstrated their latest creations to an invited audience—after quietly securing a patent. And even the devoted Curies were not above parlor tricks. Guests at their home were challenged to see a test tube of glowing radium—through closed eyes!

Today, science fairs enjoy great popularity among audiences and participants alike. That's why we find many types of science fairs, from hall and classroom displays in elementary schools to high school competitions at state and national levels where each participant researches, designs, and carries out an original study. In addition, many corporate-funded fairs and competitions offer sizeable scholarship rewards.

This book contains a variety of science fair projects, including electricity, physics, optics, aerodynamics, biology, ecology, chemistry, meteorology, astronomy, geology, and perception. These projects not only amaze, entertain, and educate, but they also clearly demonstrate a variety of scientific principles in new and exciting ways. Attractive model-making is emphasized—projects look good on display, operate smoothly, and engage audiences and science-fair judges alike. The projects demonstrate sophisticated scientific principles—a freestanding arch constructed of specially designed interlocking blocks, a giant bubble-making machine, and a stereoscopic viewer with its own set of 3-D pictures. Classic projects have a new twist—a sound dish allows you to hear distant conversations, a telescope is adapted for sunspot observation, and a many-layered crystal shimmers with brilliant colors. Timely projects explore evolving fields of study, like the effects of acid rain on marble and the mysterious workings of insect communication.

Each project contains easy-to-follow instructions and clear illustrations to aid each stage of construction. In most cases, you'll only need ordinary household objects; tools found in a toolbox, basement, or garage; or access to a hot plate or stove. Other projects

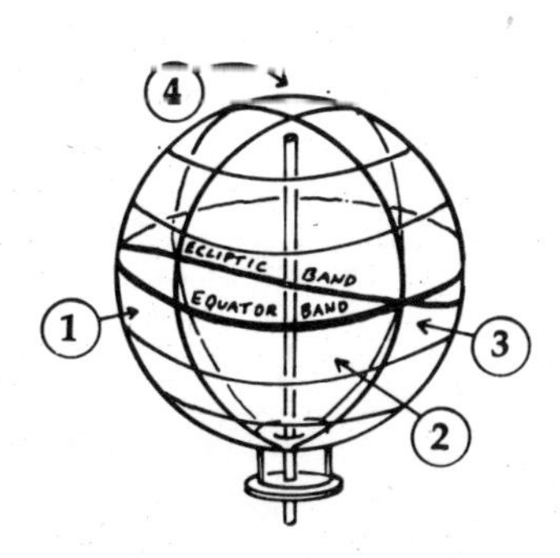

may require supplies and services from a science supply house, lumberyard, hardware store, arts and crafts supplier, pharmacy, electronics store, and plastic or industrial manufacturer. You may also find what you need at flea markets and garage sales. Your teacher may have some ideas, too.

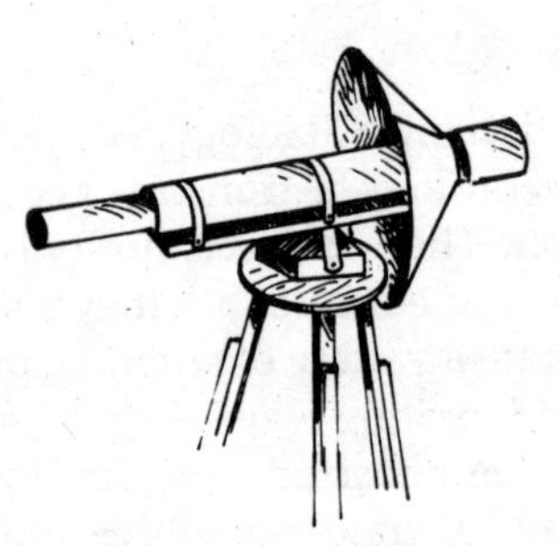

Most projects you can make yourself, but some require materials and procedures (depending on your skill and age) best done under adult supervision. When anything looks difficult, ask for help. Dry ice, corrosive or volatile chemicals, quick-bonding glues, and hot plates all require special caution. Also, be careful (especially if you've never had workshop) handling saws, screwdrivers, drills, hammers, and anything pointed, sharp-edged, or breakable like glass.

Lists of materials and hardware dimensions, with rounded-off metric equivalents (1 inch = 2.5 cm), indicate how the model shown was constructed. But you can devise your own variations and work with materials available to you. Design details matter less than your comprehension of scientific principles behind the designs. When you know the science, you can add your own unique touch to a tested formula.

And, to help you along that road, many projects begin with a concise, up-to-date explanation of scientific theories. Before duplicating Faraday's magnetic coil, you'll learn about alternating current; before constructing a polarized light box, you'll understand relevant optical principles. Learning how a particular project works will inspire you to build it and building the project will help you understand how it works. Although you can approach each project independently, some chapters contain related

projects. So, if you wish, you can create your own weather station, complete with sling hygrometer, thermometer, barometer, weather vane, and rain gauge.

Learning, solving, creating—that's what any book of science is all about. Jump-start your imagination. Let these projects inspire you to create something new, different, grander, stranger. Once you know the basic science, you'll soon get bright ideas of your own.

Electricity, the Invisible Force

Electroscope
Testing Conductivity
Wet & Dry Batteries
Juice from a Lemon
Copperplating
Circuits, Lights & Buzzers
Electric Circuit Puzzles
Working Fuse
Glowing Light Bulb
Electromagnets
Electromagnetic Crane
Homemade Galvanometer
Faraday's Discovery
Solar Cell
Wind Turbine

Electroscope

You Will Need

- Small glass jar about 6 inches (15 cm) high
- Wire coat hanger
- Thin ply aluminum foil
- ½ × 30-inch (1.25 × 7.5-cm) strip of aluminized Mylar film from a helium balloon
- 12-inch (30-cm) dowel
- Rubber balloon
- Flat transparent plate *or* glass 8 inches (20 cm) square
- Aluminum pie dish
- Cardboard
- Thin tissue paper
- Electrical *or* masking tape
- Plastic comb
- Wire clippers
- Carpenter's glue

Static Electricity

Invisible as the air, **static electricity** surrounds us. When you comb your hair on a dry day, touch a doorknob after walking across the carpet, hear the snap of sparks in your sweater, or marvel at lightning flashing through the clouds, you're experiencing freely flowing, unharnessed electrical power.

Friction—something rubbing against something else—creates static activity. A comb rubbing against hair, carpet against feet, or strands of a sweater against themselves (like ice crystals in a cloud) all produce a force called a **charge,** which either *attracts* when exposed to the opposite charge or *repels* when exposed to the same charge.

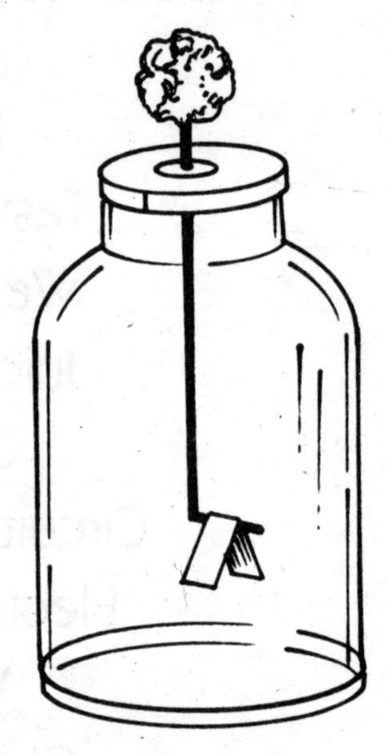

The many methods of measuring and creating static electricity in the lab provide scientists with information concerning the nature of matter.

Making the Electroscope

You can build an **electroscope,** a simple device for studying a static charge. The design here is basic, but once you understand the concept you'll be able to construct instruments large or small, plain or fancy, to suit your needs.

Lay out the parts of your electroscope on a surface with plenty of room to work. Use wire clippers to cut off a straight piece of coat hanger about 5 inches (12.5 cm). With pliers, bend a 2-inch (5-cm) section at one end into a right angle.

Turn the jar upside down onto the cardboard, and trace a circle around the opening. Cut out the circle and punch a hole in the center with your pencil. Now, carefully push the wire through the hole, straight end first, so that only about 2 inches (5 cm) of wire protrudes through the opposite side of the circle.

With as little tape as possible, attach the middle of the Mylar strip to the bent end of the wire, so that the strip hangs down in two equal halves. Glue the cardboard circle to the top of the jar, with the bent end of the wire with the Mylar strip facing down. Tape the edges where the circle touches the rim of the jar, and place a small band of glue around the wire where it punches through the cardboard. These precautions will prevent the electrical charge from leaking.

After the glue dries, crumple the aluminum foil into a tight ball about ½ inch (1.25 cm) in diameter, and carefully push it onto the top of the wire.

Testing the Electroscope

Let's test the electroscope. Rub the plastic comb against your hair, a wool sweater, or a fur-like material, and hold it close to the aluminum foil ball. The ends of the Mylar strip inside the jar fly apart! If the humidity is high, you may not get dramatic results. Try warming the electroscope in an oven for a few minutes to dry any moisture; then try again. Your electroscope should work well in any air-conditioned room or in a heated room in winter.

Positive & Negative Charges

What does an electroscope demonstrate? Besides the presence of static electricity, an electroscope shows the attraction and repulsion of electrical charges. When you rub the comb, friction causes a *positive charge* to build up in the plastic. When you hold the positively charged comb near the aluminum foil ball, it attracts *negative charges* which move up through the wire so that only positive charges remain in the Mylar strip. Both ends of the strip now have the same charge, so they fly apart!

Try substituting different objects for the plastic comb, arranging them neatly beside your electroscope. Invite people who view your exhibit to experiment and test the electroscope themselves!

Other Toys

You can also make a few toys that operate on this static electricity principle and display them beside your electroscope. An electrostatic airplane consists of a thin piece of aluminum foil (Mylar is too flimsy), cut into the shape of a small airplane, and an inflated balloon mounted at the tip of a 12-inch (30-cm) dowel.

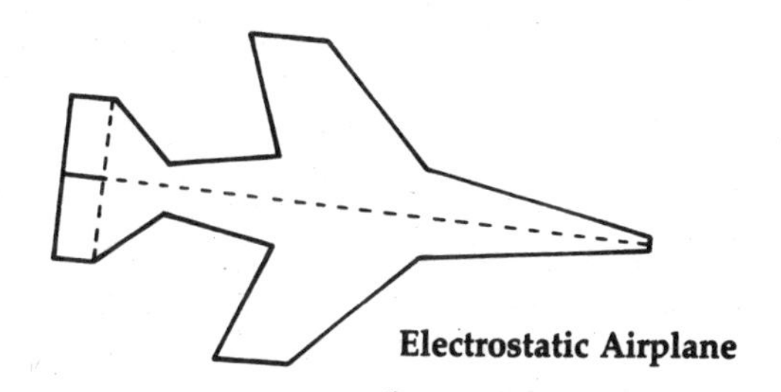

Electrostatic Airplane

Cut the foil airplane pattern, and fold the plane along the dotted line. Then unfold it, but do not flatten it completely. Fold the tail flaps up along the dotted line. Attach a few squares of tape to the nose to improve the flight.

Inflate the balloon and stretch the neck over the tip of the dowel. Wrap tape around the end of the dowel to ensure a tight fit, as necessary.

Rub wool or fur against the balloon to give it a negative charge. Toss the airplane in the air and touch it with the balloon. Now both airplane and balloon have negative charges and will repel each other. Use the balloon to keep your airplane in the air by repulsion. If you find that the balloon-on-a-dowel is too unwieldy, try substituting a plastic baseball bat or any other plastic rod or wand. Rub it vigorously with wool or fur as before.

Another toy involves tissue paper figures that jump and dance in an aluminum pie plate. Cut the figures just a little shorter than the depth of the pan. Place them in the pan, and cover the pan with a transparent dish or piece of glass. Rub the top surface with a piece of wool or fur, and the figures will spring to life!

Testing Conductivity

You Will Need

- 3 inches of insulated copper wire, cut into 3 equal sections
- Small cork
- 5 nails or brads, × 1½ inches long
- Sample conductors—aluminum foil, soapless steel wool, microscope slide, wooden stick, charcoal stick, graphite pencil, plastic ruler, small piece of rubber tubing
- 1 tablespoon of salt
- 1 tablespoon of baking soda
- 1 tablespoon of granulated sugar
- Small glass bowl
- ¼-inch (.62-cm) plywood 4 × 4 inches (10 × 10 cm)
- 20 alligator clips for wires
- 4.5-volt battery
- Flashlight bulb
- Carpenter's glue
- 6 feet of insulated wire
- ¼-inch (.62-cm) plywood 12 × 18 inches (30 × 45 cm)
- Piece of cloth
- Plant leaf (alive)
- 2 strips of 1 × 2 wood

Electricity Conductors

Rubbing two materials together (static electricity) creates an electrical charge which can be transferred from one object or person to another. But how does an electrical charge travel? Electricity, like a stream of water, flows through many substances—solid, liquid, and gaseous. Lightning flows through moist air, and a "carpet shock" flows through your body.

A substance that allows electricity to flow through it easily is called a **conductor.** A substance that allows only a partial flow of electricity is a **resistor.** And a substance that rejects the flow of electricity completely is called an **insulator.** All three types of substances are often used in combination when electricity is at work.

Why some substances conduct electricity and others do not once puzzled scientists. But they found the answer with an understanding of **atoms** and how they attract each other. Good conductors, like metals, are composed of atoms with freely moving **electrons.** That means the atom's electrons do not remain in fixed orbits around the **nucleus;** instead, they move freely in a fluid state which allows an electric current to pass through easily. Still, a strong attachment, called a **metallic bond,** exists between these atoms.

Nonconducting materials are held together by **covalent bonds**. Electrons in these substances are attached to the nuclei of atoms in a fixed position and cannot move about freely. Visualize a metallic bond as a kind of bead curtain and a covalent bond as a chain-link fence.

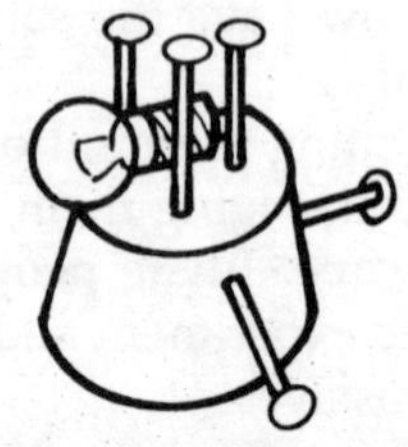

Light Bulb Indicator

Testing, Testing

Let's test various solid substances for conductivity, using a light bulb indicator.

To make the indicator, use the cork and secure a flashlight bulb with three 1½-inch (3.25-cm) nails. For terminals, drive the two remaining nails into the cork's side so that they touch two of the vertical nails.

Battery

Next, we'll need a battery to test our indicators, and a snappy mounting board for the battery, something usable in future projects. Position the plywood horizontally on a flat surface near the bottom, draw a 4 × 12-inch (10 × 30-cm) rectangle. Paint this rectangle with diagonal black and yellow stripes.

Center the 4 × 4-inch (10 × 10-cm) plywood battery brace along the board's right edge, and nail it in place.

Mounting Board with Battery

Place a thumbtack at each of the four corners of the plywood, and glue a fifth thumbtack upside-down near the middle of the board to hold the light bulb indicator. Cut the wire into one long and two short sections, and attach alligator clips at the ends.

Press the light bulb indicator onto its tack, attach the battery to the brace with a rubber band, and loop the wire around the tacks (see diagram).

Test the simple circuit (closed system of electrical flow) by attaching one wire from the battery to the horizontal nail terminal of the bulb indicator and the other wire to the vertical terminal. If the bulb lights, all is well. If not, reposition the nail terminals in the cork.

Draw a chart to record results of your conductivity tests. Then begin testing with the microscope slide. Place it in the testing area and clip the wires to opposite ends of the slide. Does the indicator bulb light? Note your results on the chart. Next, try the aluminum foil and the remaining substances, noting your results. In some cases, the bulb will light but only weakly. Rate conductivity on a scale of "excellent" to "poor."

Record results after each test. For a conductor of electricity, the bulb lights up. Does a conducting substance with a metallic bond always have the properties of a metal? What about the graphite pencil "lead"? Graphite is a conducting form of carbon, but the carbon in charcoal is nonconducting. That's because carbon is a substance that can conduct or resist, depending on its form.

Make several blank charts and invite viewers to participate. They can test small objects they may be carrying for conductivity, such as key chains, pocket combs, hair clips, and mirrors.

Ionic Bonding & Electrolytes

When a metallic element combines with a nonmetallic element to form a compound, a different type of bonding—ionic bonding—occurs. In this type of bond, atoms in the metal lose electrons, which are acquired by the atoms of the nonmetal. The charged particles formed from this exchange are called **ions,** and they remain in a fixed position, like the electrons in a covalent bond. However, when certain compounds are melted or dissolved in water, the ions separate and move freely, like the electrons in a metallic bond. The resulting liquid is called an **electrolyte.**

We'll demonstrate the conductivity of some compounds in both solid and liquid form. Draw another data chart to record results.

Testing Electrolytes

Pour the 2 tablespoons of salt on the index card and test it for conductivity with your circuit. Next, dissolve the salt in a small bowl of water. Test again for conductivity and record the result. Although the salt was a nonconductor in the solid state, it easily conducts electricity in a liquid—or electrolyte—state. This is also true for baking soda. Both substances have ionic bonds and conduct electricity only when in liquid form.

Repeat the test for sugar. Sugar does not conduct electricity in the solid or in the liquid state. The compound of sugar is held together by covalent, *not* ionic, bonds.

Some electrolytes exist only as dissolved or melted forms of solid compounds. However, some acidic liquids, such as vinegar or lemon juice, are natural electrolytes. Water, too, easily conducts electric current, particularly when it is "hard" (filled with dissolved minerals).

Assemble a collection of additional compounds that dissolve easily in water, like cornstarch, bicarbonate of soda, and Epsom salt, and invite viewers to test them. The results may surprise you!

Substance	Conductivity as a Solid	Conductivity in Solution	Electrolyte or Nonelectrolyte	Bond Type
Sugar				
Table salt				
Baking soda				
Cornstarch				
Epsom salt				

Wet & Dry Batteries

You Will Need

- Wide glass jar or beaker
- Medium-size bowl
- Pint (480 ml) of vinegar
- 4 to 8 ounces (120 to 240 ml) of ammonium chloride
- 2 strips of zinc *or* galvanized sheet metal 1 × 5 inches (2.5 × 12.5 cm) and 1 × 6½ inches (2.5 × 15.6 cm)
- 2 strips of copper
- Strip of thin felt ¾ × 6 inches (1.8 × 15 cm)
- 2 LED lamps
- 30 inches (7.5 cm) of insulated copper wire
- Commercial D-size dry cell battery (used up)
- Metal shears
- Small 17-gauge or narrower brad
- Steel wool or fine-gauge sandpaper

Battery Basics

Our daily lives wouldn't be the same without batteries. From flashlights to cars, they supply portable energy. Here we demonstrate how batteries work, using an **electrolyte.**

Since our homemade battery produces only a weak current, a low-voltage Light Emitting Diode (LED) lamp will serve as our indicator. LEDs have wire terminals attached to the base—and are used in calculators and wristwatches to light up the numbers. They use very little energy and give off a bright light.

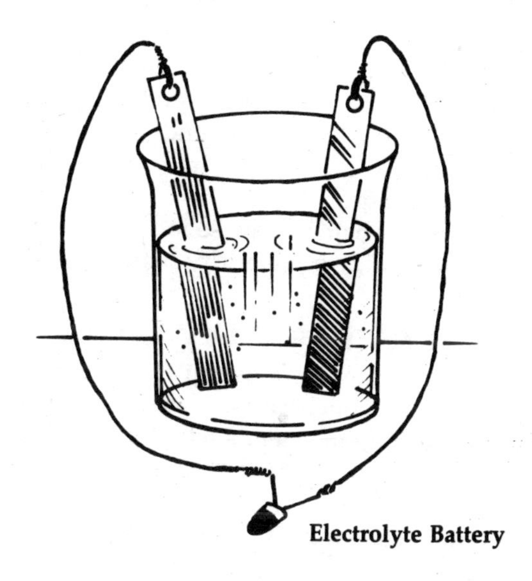

Electrolyte Battery

Constructing a Wet Battery

Fill the glass with vinegar. Vinegar is an acetic acid that conducts very well—in other words, it's an electrolyte. Cut the copper wire into four lengths: two 10-inch pieces (25 cm) and two 5-inch pieces (12.5 cm). Use scissors to strip insulation off the end wires' ends.

With the brad and hammer, punch a hole in the top of the 1 × 5-inch (2.5 × 12.5-cm) zinc and copper strips. Then, loop one end of each wire to the copper and zinc, respectively. Twist the opposite ends of the wires around the leads of the LED lamp. Carefully dip the two pieces of metal (the electrodes) into the vinegar. The lamp should begin to glow. If nothing happens, reverse connections on the terminals.

The glowing lamp indicates an electrochemical reaction occurring in the vinegar which leaves an excess of *negatively charged electrons* (ions) in the zinc. This creates a small current that flows through the wire to the copper strip (seen in the LED's glow). The copper strip, in turn, pulls *positively charged electrons* from the acid and itself becomes positively charged.

Other Wet Batteries

Many variations in wet battery construction are possible by combining different electrodes and electrolytes. Lemon juice, tomato juice, vinegar and salt, and compounds such as copper sulfate and ammonium chloride (sal ammoniac) make excellent electrolytes. With each of these liquids, different pairs of metals, such as tin and nickel, silver and iron, or tin and aluminum, make efficient electrodes.

Since wet batteries are so inconvenient to carry around, we prefer dry batteries.

Constructing a Dry Battery

For a dry battery, we use an electrolyte and more metal—alternating sheets of copper and zinc with moistened felt. This is a copy of Alessandro Volta's 1880 forerunner of the modern dry cell. You'll also need the ammonium chloride.

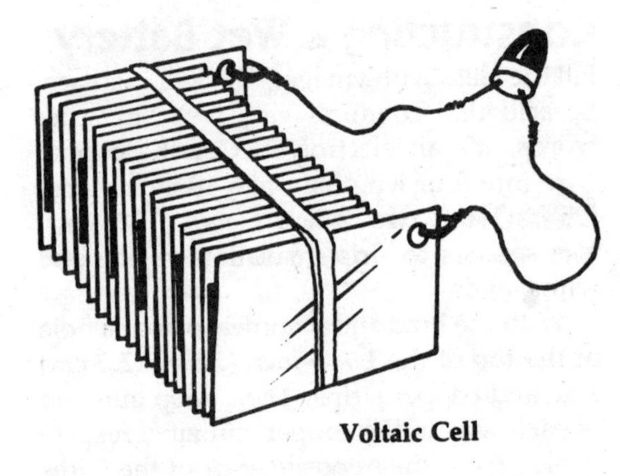

Voltaic Cell

With sandpaper or steel wool, clean both sides of the 1 × 6¼-inch (2.5 × 15.6-cm) copper and zinc strips. Then put both strips into five 1-inch (2.5-cm) squares. The sixth square will be a little longer than the rest. Hammer the squares so that the edges are as flat as possible for tight connections between them. Cut the felt into six pieces ¾ × ¾ inches (1.8 × 1.8 cm).

For the electrolyte, mix enough ammonium chloride with a little water in a bowl, to wet all six pieces of felt. Soak the felt in the solution while you work on the next step.

Use the brad and hammer to make a small hole in the narrow edge of the longer copper piece. Do the same for the longer zinc piece. Strip one end of two copper wires, and loop each end of the wires through the holes you just made. Connect the other end of each wire to the LED terminals.

Begin assembling the battery by placing the long copper piece (with the attached wire) on a flat surface. Remove a felt square from the electrolyte and squeeze out the excess fluid. Place this moist square on top of the copper, and place one of the smaller zinc squares on top of the felt square. Another piece of wrung-out felt goes on top of that, followed by a smaller copper square, and so on, until you reach the longer piece of zinc with wire attached.

Secure the metal and felt squares together with a rubber band. The LED lamp will glow until the electrolyte dries up. Then, gently replenish the liquid by poking an eye dropper between the layers.

Your Exhibit

For your exhibit, display your homemade wet and dry batteries next to a commercial D-size battery. Draw a diagram of the inside of a commercial battery, noting the similarities with your vinegar battery. We've come a long way!

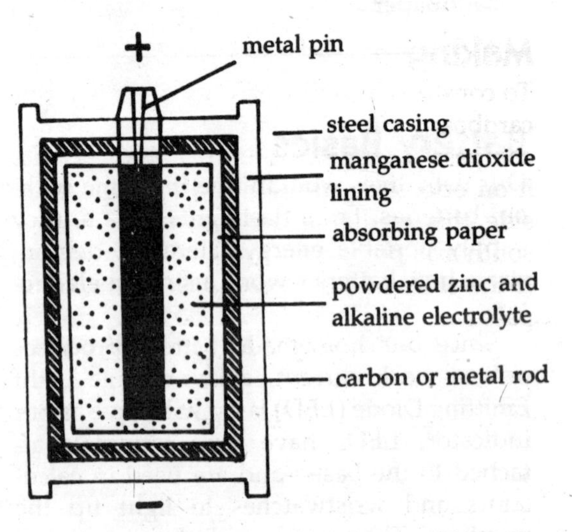

Cross Section of Alkaline Battery

Juice from a Lemon

You Will Need

- Lemon
- Strip of zinc or galvanized sheet metal, ½ × 3 inches (1.25 × 7.5 cm)
- Strip of copper, same size
- 5 feet of insulated copper wire
- Small compass

Lemon Power

The citric acid in lemon juice provides an excellent conducting environment—or **electrolyte**—for electric current. Perforated with a strip of zinc and a tube of copper, a whole lemon produces an electron flow in a closed circuit. In other words, it becomes a battery!

In this case, however, the current produced is too minuscule to light even the smallest bulb. To detect the electricity in a lemon battery, we must use a compass, turning it into a **galvanometer,** or electron sensor.

Making a Galvanometer

To construct a simple galvanometer, cut the cardboard into a rectangle, as wide as the compass but long enough to fold up around it on opposite sides. Place the compass inside this cradle, making sure the north–south axis points towards the folded sides. If the sides of the cradle obscure the compass face, trim the cardboard. Next, stretch out 5 inches (12.5 cm) of copper wire, and wrap the remaining wire around both the compass and cradle directly over the north–south axis. A hundred twists of wire should be sufficient, leaving about 5 inches (12.5 cm) of wire at the end.

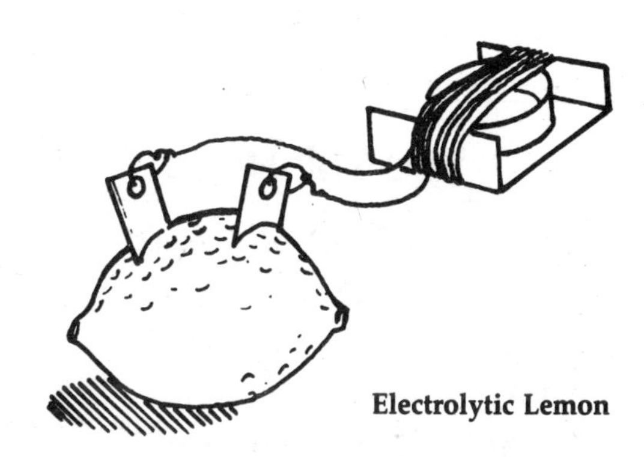

Electrolytic Lemon

Now, strip 1 inch (2.5 cm) of insulation from the ends of the coiled wire, and tape each end to the zinc and copper tube respectively. Your galvanometer is complete.

Lemon "Juice" Test

To prepare the lemon, gently crush it in the palms of your hands so that the juices flow freely inside. Cut two small slits at the top of the lemon, no more than 1 inch (2.5 cm) apart. Gently insert the zinc strip into one slit and the copper into the other.

Now, watch the galvanometer. The needle should begin to swing from a north–south direction to an east–west direction, indicating a flow of current. This occurs because electricity in the coil creates a weak **magnetic field** around the compass, swinging the needle from its natural position.

Copperplating

You Will Need

- Brass key or small brass object
- 6-inch (15-cm) strip of copper
- Plastic charm
- Tube of powdered graphite
- Plastic gallon milk container
- Craft knife
- 4.5-volt battery
- 2 feet (60 cm) of insulated copper wire, cut into equal lengths
- Carpenter's glue
- Dishwashing detergent
- Vinegar
- Table salt

Making a Copperplate Conductor

Vinegar and salt water each make an excellent electrolyte solution. When vinegar and salt are combined, they create an even more conductive brew that's crucial to the success of this project.

With the craft knife, carefully cut the plastic milk container down into a 4-inch- (10-cm-) deep tray. Fill the tray with just enough vinegar to cover the key as though it were standing on end. Add salt. Stir until the solution becomes saturated and no more salt dissolves.

Tape or solder one piece of wire to the positive terminal (the side with the bump) of the battery. Then, tape or solder the other end of the wire to the strip of copper. Carefully bend the copper so that it clips onto the side of the tray with at least 3 inches (7.5 cm) of it hanging into the vinegar–salt solution.

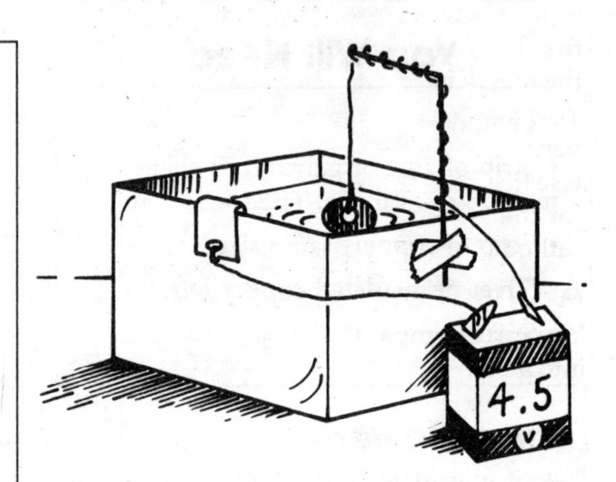

Copperplating Setup

Wash the brass key with dishwashing detergent and water to remove dirt and grease. Loop one end of the second piece of wire through the hole of the key, twisting it securely. Connect the other end of the wire to the negative terminal of the battery.

Straighten the paper clip, and tape the tip to the outer edge of the tray. Carefully bend the clip about 45 degrees over the solution so that it resembles a fishing rod. Hang the key end of the wire from this rod so that the key is submerged in the solution. Loop the wire around the paper clip once or twice to keep it from slipping, as needed.

Electroplating Process

Watch as bubbles form on the key and the electrolyte solution begins to change color. While **electroplating** proceeds, hydrogen is released from the water in the vinegar, creating bubbles. If too many bubbles accumulate on the surface of the key, remove it from the solution just long enough to wipe it clean. After about 20 minutes, the key begins to assume a copper hue. In an hour, it should be thoroughly plated.

The more active metal—in this case, copper—loses electrons to the less active metal, brass. Here, an electric current supplied by the battery speeds up the process.

In electroplating operations, the metal that loses electrons is called the **anode** and the metal that gains electrons is the **cathode.** You may also electroplate nickel over copper, since nickel is the more active metal. Use the same setup, but with a strip of nickel for the anode and a strip of copper for the cathode.

A Charming Solution

You can also turn an ordinary plastic charm into an attractive piece of jewelry by copperplating it. Of course, plastic is a nonmetallic substance, so it must be coated with a metallic substance to attract the copper electrons. Powdered graphite—a metallic form of carbon—provides the answer.

Begin by dipping the charm into a small cup of watered-down carpenter's glue. Let the glue dry until it feels tacky; then, sprinkle the graphite over the charm until you coat it completely. Allow the coated charm to dry overnight, if necessary.

When it's thoroughly dry, suspend the charm from the paper clip and dip it in the vinegar–salt electrolyte, just as you did the brass key. Connect the wires to the battery, and watch the ordinary charm turn into something lovely.

Circuits, Lights & Buzzers

You Will Need

- Light bulb indicators (see light bulb indicator, p. 10)
- Spool of insulated copper wire
- Spool of uninsulated (enameled) wire
- 3 tin soup cans *or* 1 coffee can
- Wood blocks
- Flashlight batteries
- Electrical tape
- ¼-inch (.62-cm) thick plywood 1½ × 1 feet (45 × 30 cm) (for base)
- 2 pieces of ¼-inch (.62-cm) thick plywood 2 × 3 inches (5 × 7.5 cm) and 3×6 inches (7.5×15 cm)
- 3-inch (7.5-cm) piece of steel hacksaw blade
- L-shape iron or brass bracket with two eyeholes and one arm at least 1½ inches (3.75 cm) long
- Cardboard
- 20 small wood screws
- 1½-inch (3.75-cm) long nail
- Small bolt with nut
- Thin cork sheet
- Carpenter's glue
- 4.5-volt battery
- Battery mounting board
- Metal shears

Electrical Circuit Basics

From doorbells to car ignitions, the simple electrical circuit is the foundation for the most complex electrical operations. Creating an interesting model display of the various kinds of circuit requires few tools and only an afternoon's work. If you tested the conductivity of various materials (see p. 10), you already know how to construct a simple circuit. But to operate and test these models, you'll need to make an additional part, a switch.

Constructing an Electrical Switch

You can construct as many switches as models you plan to build. Begin with an empty tin can, thoroughly washed and dried. Carefully cut off both the top and the bottom; then, cut the can lengthwise and flatten it out. From this sheet of metal cut a strip ½ × 3½ inches (1.25 × 8.75 cm) and a smaller strip 1 × ½ inch (2.5 × 1.25 cm). With the metal shears, carefully round the corners of the longer strip. *Be very careful when working with tin; the sharp edges can cut you.*

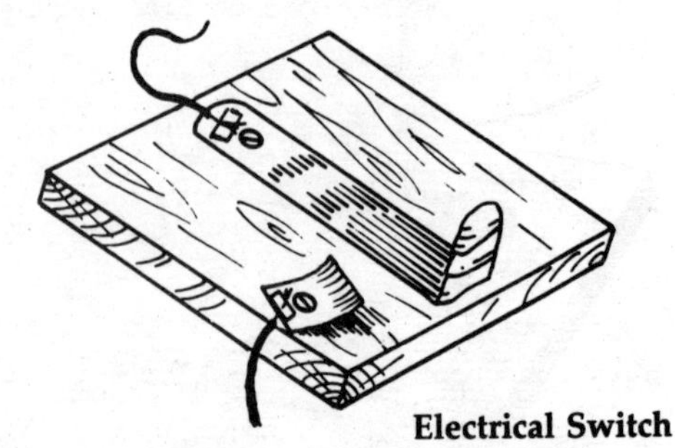

Electrical Switch

Next, cut a block of wood 3 × 3 inches (7.5 × 7.5 cm), and draw a line down the middle, dividing it into two equal halves. Place the longer strip along this line so that one end protrudes about ½ inch (1.25 cm) over the edge of the wood. Attach the opposite end to the wood with a screw, making sure the entire strip pivots easily. Bend the protruding part back 90 degrees, and wrap it in electrical tape. This is where you grasp the switch.

Next, attach the smaller strip, or **contact,** to one side that's perpendicular to the free-swinging pivot, making sure the pivoting strip can partially pass under it for a good contact. Do this by screwing the smaller piece through its outside edge, and, with the tip of your finger, gently bend it back just a few degrees.

Your switch is complete, and it's in the ON, or *closed,* position when the metal strip touches the contact.

Making a Simple Circuit

Now, along with a light bulb indicator (see p. 10), you have the essential parts for your models. Well, almost.

Use the battery mounting board from the project for Testing Conductivity to display various circuit configurations.

Open your switch and place it in the testing area. Fasten the alligator clips of the free wires to the pivot and contact of the switch. Close the switch. If the contacts are intact, the bulb will light.

Doorbell Switches & Electric Buzzer

Many commonplace items, like desk lamps, radios, toys, and kitchen appliances, use a simple open/closed circuit. But some tasks require a variation of this design. A doorbell, for instance, has *two* switches, one for the front of the house and one for the back. Either switch operates the bell, and you can demonstrate this principle with the model.

Front & Back Switch

You can construct a small buzzer device without much difficulty. The buzzer operates on the principle of **electromagnetism.** Simply stated, electromagnetism is magnetic attraction created by an electric current flowing through a coiled wire.

Begin constructing the switch by cutting out a cardboard circle about 2 inches (5 cm) in diameter and pushing a small nail through the center. Hammer the nail halfway through the 2 × 3-inch (5×7.5-cm) piece of plywood.

Next, wind the uninsulated wire around the nail about 200 times, using the cardboard circle as a cap for the coil. Make sure 6 to 8 inches (15 to 20 cm) of wire hang free at each end of the coil.

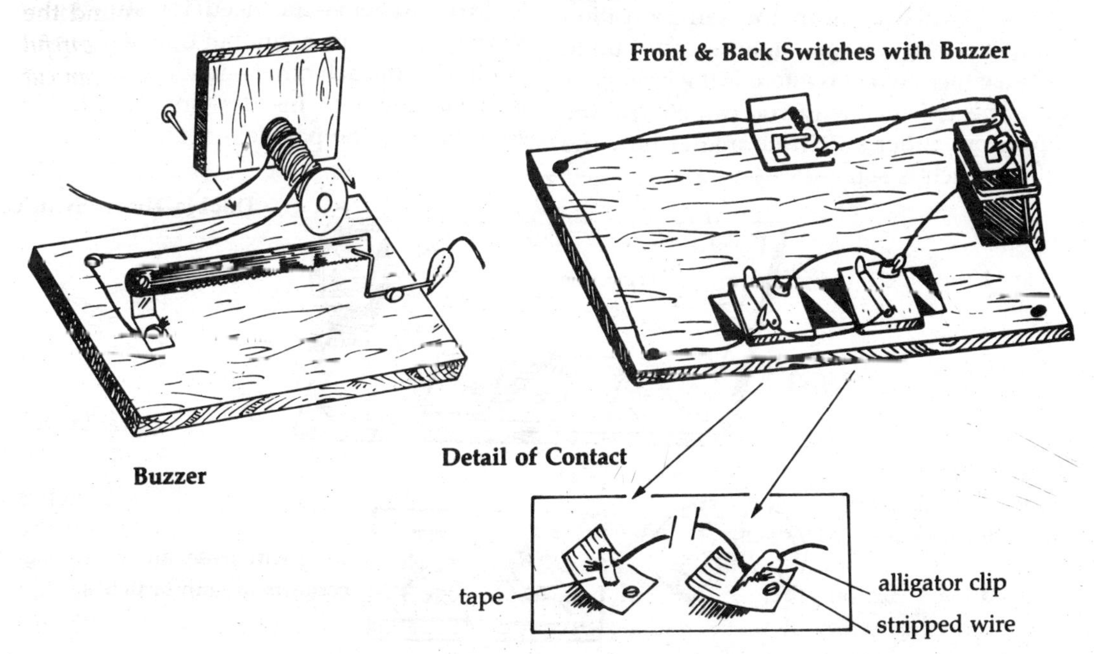

Front & Back Switches with Buzzer

Detail of Contact

Buzzer

Screw this piece to the larger piece of plywood. Bolt the 3-inch (7.5-cm) piece of hacksaw blade to the long side of the L-shape bracket, and screw the bracket into the base, so that the end of the blade extends about ¼ inch (.62 cm) past the nail head. Make sure you have at least ⅛ inch (.31 cm) between the nail head and the blade.

Bend the paper clip, and attach it to the base shown in the buzzer diagram, making sure the end of the clip barely touches the hacksaw blade. Glue a square of cork sheet to the underside of the base. Your buzzer is complete.

Connect the coil wires as shown in the diagram, and place the buzzer on the board. Connect the switches as shown. When you have completed your hookup, notice how each switch completes the doorbell circuit independently of the other.

Double-Throw Switch

A more sophisticated circuit design requires the double-throw switch, which allows someone to turn a light on or off from either the top or bottom of a staircase. To construct the double-throw switch, modify two simple switches, and combine them in the configuration shown in the Double-Throw Switch illustration.

Both switch arms in the same position complete the circuit, lighting the bulb. When either switch is altered, the light goes out. This explains why you can control the light from either switch no matter how the other switch is set.

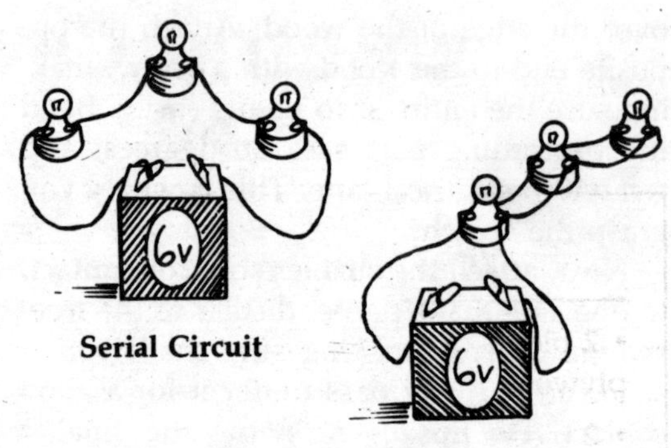

Serial Circuit

Parallel Circuit

Serial & Parallel Circuits

You might be familiar with the problem of finding the one burned-out bulb in a long string of Christmas-tree lights. It's frustrating since other bulbs won't light until you replace the dead bulb. Lights arranged in this kind of **serial circuit** depend on each other because every single bulb is an integral part of the whole circuit. If one burns out or is defective, the circuit is broken and no electric current flows. If you add more bulbs to a serial circuit, the light in each bulb grows dimmer. This is because the bulbs share and divide the total current between them, and the battery has a limited amount of power.

The challenge to find a practical solution to this problem—the need for streetlights sharing the same circuit but operating independently of each other—made Thomas Edison sit down at his drawing board. His solution was the parallel circuit.

Double-Throw Switch

Bend switch strip back to accommodate alligator clips.

Tape wire between contacts of both switches.

Electric Circuit Puzzles

You Will Need

- 2 pieces of ¼-inch (.62-cm) plywood 6 × 8 inches (15 × 20 cm)
- 2 pieces of ¼-inch (.62-cm) plywood 3×4 inches (7.5 × 10 cm) for battery brace
- Six 6-volt flashlight bulbs
- 6 enamel flashlight bulb sockets
- Spool of insulated copper wire
- 4 alligator clips
- Two 6-volt batteries (with strip terminals, if possible)
- Thin nails, no longer than ½ inch (1.25 cm)
- Thick rubber band
- Electrical tape
- Epoxy cement

First Puzzle

Constructing these puzzles will teach you more about serial and parallel circuits. You can quiz your friends, too!

For the first puzzle, position the first piece of plywood horizontally in front of you. Measure 3 inches (7.5 cm) from the top along the left and right edges and make marks there. Turn the 3 × 4-inch (7.5 × 10-cm) battery brace narrow side up, center it on the left mark, and nail it in place with the bottom edge flush with the bottom of the plywood. Repeat this for the second piece of plywood, too.

With the first piece of plywood, measure ½ inch (1.25 cm) towards the center from the mark on the right edge. Apply epoxy glue, and attach a bulb socket, making sure that the two screw contacts on the socket are parallel to the narrow edges of the plywood.

Measure 3½ inches (8.75 cm) from the top and bottom corners of the right edge and mark with pencil. Connect the marks with a vertical line. Measure 1½ inches (3.25 cm) towards the center from each mark, and glue the sockets, again making sure the screw contacts are parallel to the plywood's narrow edges.

Place a 6-volt battery against the brace, and loop a rubber band around the battery to secure it in place. Connect all wires according to the diagram, attaching alligator clips to the wire ends that will connect with

First Puzzle

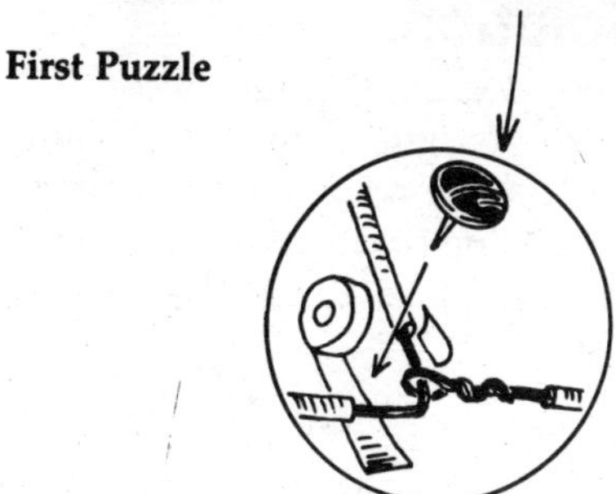

the battery. The wires should trace the plywood at a uniform distance of ½ inch (1.25 cm) from the edges. Thumbtacks help keep the wires straight and make the circuit configuration clear. Where splicing is necessary, use scissors to strip off about 1 inch (2.5 cm) of wire insulation. Then, twist the wires together and seal them with electrical tape. Use a felt-tip marker to draw letters next to the sockets (see diagram). Screw in the bulbs, but keep the battery disconnected for now.

Second Puzzle

For the second puzzle, measure 4 inches (10 cm) along the bottom edge of the second piece of plywood from the bottom right corner. Make a mark there; then, measure 1½ inches (3.75 cm) towards the center and glue a socket. Parallel to that socket and 1 inch (2.5 cm) from the right edge, glue a second socket. Between them, and 2 inches (5 cm) from the top edge, glue the third socket. Attach the battery to the brace. Use wire and thumbtacks (see diagram), making sure that the wire traces the plywood at a uniform distance of ½ inch (1.25 cm) from the edges. Screw in the bulbs, and letter each socket (see diagram).

Second Puzzle

Puzzle Solutions

Now for the puzzle part. Before you connect the battery, can you guess, just by looking at the two circuits, the relative brightness of each bulb? What kind of circuit does the first puzzle represent? How about the second puzzle?

Connect the batteries and observe the lit bulbs. A and B should be equally bright, but less bright than C; E and F are both much fainter than D.

Now try to guess how removing one bulb from each puzzle will affect the other bulbs. Start by unscrewing bulb A. Notice that bulb B goes out, but bulb C remains unaffected. Unscrew bulb F and notice the change in bulbs D and E: D remains the same while E grows brighter. Does something seem familiar in all this?

The first puzzle consists of a serial circuit (A and B) combined with a parallel circuit (C). A and B share the same amount of current, dividing it between them, which means less light from each bulb. When you remove any bulb from a serial circuit, the circuit is broken and the other bulbs go out. That's why B went out when A was unscrewed. Neither bulb affected C, however, since that circuit remained intact.

The second puzzle is also a serial circuit, but it is branched so that the circuit can reroute if one bulb goes out. Placing E and F so that they branch out from D means that E and F split the current from D, itself reduced by being in a serial arrangement. When you unscrew F, bulbs E and D now share equal parts of the current and should appear as bright as bulbs A and B of the first puzzle.

You can predict other results as you consider removing and replacing bulbs. It's a great way to review basic circuitry concepts. Enjoy!

Working Fuse

You Will Need

- 4 feet (120 cm) of insulated copper wire (not too thin), cut into various lengths
- 6-volt battery
- Light bulb indicator (see light bulb indicator, p. 10)
- Simple switch (see p. 18)
- Strands of Christmas-tree tinsel cut into ½-inch (1.25-cm) lengths
- 8 alligator clips
- Battery mounting board
- Piece of cork sheet
- Craft knife
- 2 notched mounts, cut from cardboard

Fail Safe for Overloaded Circuits

Thomas Edison recognized the need for an invention to keep electrical circuits from overheating. His solution was the electric fuse, patented under the name "safety conductor for electric lights."

Like an automatic safety switch, a fuse cuts off electric current when it becomes strong enough to cause a fire. Dangerous levels of electric current occur when a main line overloads because too many branch circuits are in use at the same time. For example, operating several household appliances at once often leads to a situation where the fuse must do its job. A power failure, due to a blown fuse, is usually the result.

The fuse, placed along the main circuit wire, melts from the heat of too much current, thereby breaking the circuit so that all electrical activity stops. That's why the materials used for fuses must be good conductors but have a *low melting point*. We'll use a narrow strip of aluminum Mylar or Christmas-tree tinsel for the fuse.

Constructing a Simple Fuse

Use the battery mounting board from the Testing Conductivity project. Modify it by placing a light bulb indicator on the left side of the board, and by attaching a new component, the fuse holder.

Construct the fuse holder from cardboard strips, notched at the top, and inserted into a strip of cork sheet. Use a craft knife to slice grooves in the cork for inserting the cardboard. Place the fuse holder over the thumbtack at the top of the board.

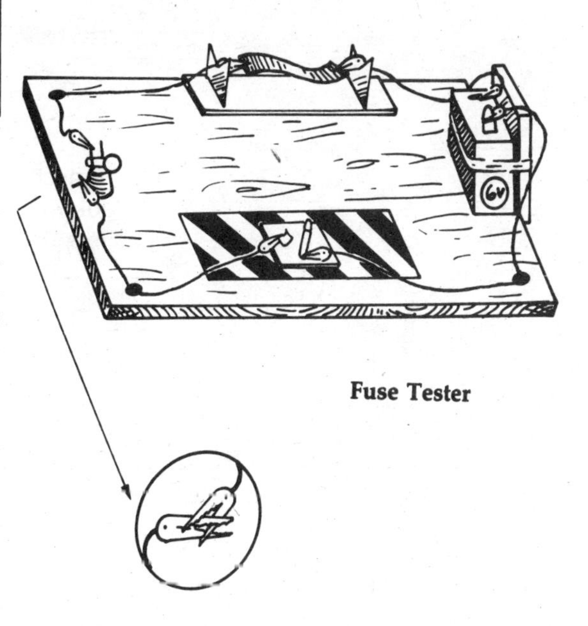

Fuse Tester

Connect everything according to the diagram, placing alligator clips at the ends of wires where appropriate. Depending on the thickness of your copper wire, the resistance of the switch may keep your fuse from melting. If this appears to be the case after a few tryouts, disconnect the switch. Close the circuit and observe the bulb. If the bulb fails to light, test all your connections.

Now you want to create a situation where the circuit overloads and the fuse does its work. Remove the alligator clips from the bulb indicator, and clip the two free ends of wire together. Without the bulb to act as resistance, the electric current becomes much stronger, heating the strand of tinsel until it melts and breaks the circuit. Watch closely, because this happens only seconds after you clip the bulb wires together.

Make sure you have a good supply of tinsel around so that observers may test your fuse model themselves.

Glowing Light Bulb

You Will Need

- Switch (see Constructing an Electrical Switch)
- Small glass bottle 3 × 4 inches (7.5 × 10 cm) or 250-ml glass flask
- Cork or rubber stopper
- Craft knife
- 3 feet (90 cm) of insulated copper wire, cut into 3 equal lengths
- 12-volt lantern battery
- 1-inch (2.5-cm) piece of cable (braided)
- Thin nail
- Steel wool

A Matter of Substance

Among the three substances—conductors, resistors, and insulators—the **resistor** most effectively turns electric current into light and heat. An electric light bulb consists of a **filament,** which acts as the resistor, mounted in a **vacuum chamber.** When electric current passes through this filament, the resistance builds up heat until it becomes incandescent.

Our light bulb uses a single strand of cable picture-hanging wire for the filament, but another type of filament will be tested later.

Turning On a Light Bulb

Cut the copper wire into three 1-foot (30-cm) sections. Strip a little insulation from both ends of the first section, but for each remaining section, strip a full 2 inches (5 cm) off one end. The filament will straddle these 2-inch (5-cm) sections of exposed wire.

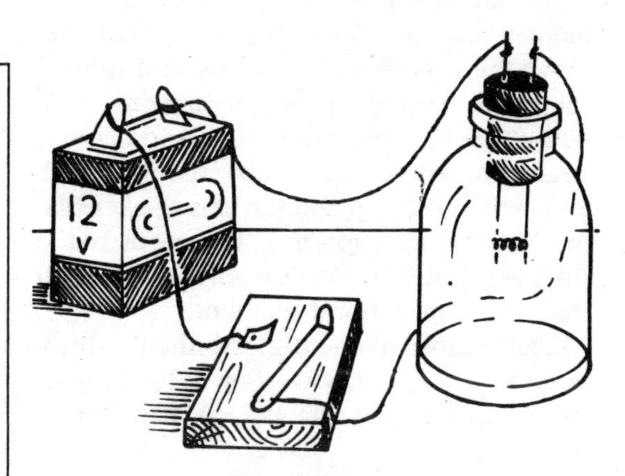

Light-Bulb Setup

Cut down the cork with a craft knife until it fits snugly into the bottle's opening. Remove the cork from the bottle, and with the nail, carefully poke two holes through it, just wide enough to pass through the stripped wire ends.

Pluck a single strand from the picture-hanging cable, and wrap it tightly around the pencil. Carefully slip off the pencil so that the wire retains its coiled shape. Twist the ends of the filament to the exposed copper wire protruding from the cork. Finally, insert the cork in the bottle with the filament pointing down.

Attach the free end of wire protruding from the cork to one terminal of the battery. Attach the free end of the second wire protruding from the cork to the contact plate of the switch; *make sure the switch is in the* OFF *position.* Attach one free end of the third wire to the switch lever and the other to the remaining battery terminal. Your completed light bulb now waits for the first test.

Through Thick & Thin

Turn the switch to the ON position and watch the filament in the glass bottle. When your light bulb begins to glow, switch it off. Wait a few moments, then try again. If you don't allow it to burn long, your light bulb can be switched on and off several time before the hot filament combines with the oxygen in the bottle and burns out. That's why commercial light bulbs contain no air.

Switch off your light bulb, and wait a few moments for the filament to cool. Then, remove the cork from the bottle, and replace the single strand of picture-hanging wire with three strands of the same kind of wire, twisted together. Reassemble the light bulb and switch it on. Although this thicker filament also glows, you might be surprised to discover that it is considerably *less* brilliant than the single-strand filament.

The thinner filament glows the brightest because it has a *higher resistance* to the electrical current than the thicker filament. If you think of the filament as a pipe and electric current as a steady flow of water, the water squeezes and accelerates while passing through a thinner pipe, much more than it would through a thicker one.

For the final test, replace the thick wire filament with a piece of steel wool. Switch on the light bulb and observe what happens. The steel wool glows for a short time, then disappears. It behaves like a **fuse,** melting and breaking the circuit when too much current passes through it.

Can you think of another device that operates like a light bulb but produces more heat than light? It warms large rooms on winter nights: the electric coil heater.

Electromagnets

You Will Need

- 2 iron bolts
- 15 feet (4.6 m) of uninsulated (enameled) copper wire, cut into 3 equal lengths
- Flashlight battery
- Transparent tape
- Small metal objects
- Small compass

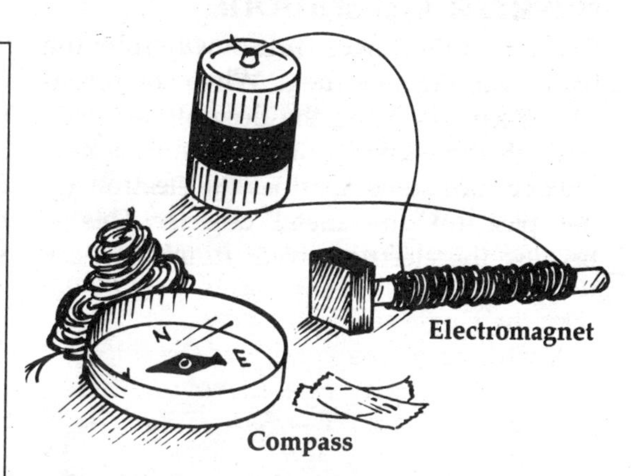

Magnets & Electromagnets

Magnets are amazing and mysterious. Some substances, like iron ore, possess natural magnetic properties. Other substances become magnetic only when electricity is added; they are called **electromagnets.** With just a flashlight battery, wire, and iron nails, you can construct electromagnets of surprising strength. Use a compass to test and compare the **magnetic field** of your electromagnets.

Single-Coil Electromagnet

Cut the wire into three 5-foot (150-cm) lengths. Wrap one around the nail about 50 times, leaving 6 inches (15 cm) of loose wire at both ends to connect to the battery. Tape the free ends of wire to opposite sides of the battery.

With electromagnets, the closer the coils of wire, the stronger the magnetic force. Also notice that wrapping the wire in one direction ensures that the electric current flows in one direction. This **single-coil** design creates a magnetic field that converges at the tips of the nail. We call these points of convergence **magnetic poles**—north and south.

Compass Tests

Place the electromagnet's tip against the side of the compass, perpendicular to the natural north–south orientation of the needle. Notice how the compass needle swings 45 degrees—repelled by the electromagnet—until it stops and points west. Positioning the electromagnet against the compass causes the needle to align itself to the magnet's north–south polarity. Place the electromagnet's opposite tip against the compass. The needle is attracted 45 degrees until it stops directly in front of the tip. Again, the compass reflects the north–south polarity of the electromagnet, now reversed.

Remove the compass, and test the electromagnet's strength by holding it against small metal objects. Disconnect one wire from the battery and notice how the magnetic attraction stops.

Double-Coil Electromagnet

For the **double-coil electromagnet,** wrap the wire around the nail as before, leaving 6 inches (15 cm) at the ends. Then, wrap the second wire around the nail, but leave about 1 foot (30 cm) of wire free at the ends. Connect the free ends of both coils to the battery's ends, and hold the electromagnet against the compass. The compass swings as before, but with a bit more force. Test the electromagnet's strength by holding it against the small metal objects. The double-coil design creates considerably more magnetic attraction.

Reverse Operations

Disconnect the longer ends of wire from the battery, and reverse them. When you reconnect them, reversing the flow of electricity through the second coil, the compass needle does not move nor does the electromagnet pick up any metal objects. This is because the electric current running in opposite directions through the coils creates magnetic polarities that balance and neutralize each other. In other words, the electromagnet no longer attracts.

Through models like these, Thomas Edison found he could vary the degree of magnetism, change the direction of current flow through coils, and change the amount of flow. He found a way to combine these functions across a single wire, calling it the **duplex system.** This was useful in his designs for the telegraph.

Electromagnetic Crane

You Will Need

- 4 round oatmeal boxes (18-ounce or 540-ml size)
- Shoe box with lid
- Narrow wafer mint box (bottom half)
- Tube from gift wrap
- 2 spools
- Corrugated cardboard
- 2 wooden dowels (or pencils) to fit through spools
- 4 triangular grip tubes for pencils
- Craft knife
- Iron bolt
- 20 feet (6.1 m) of uninsulated copper wire, cut into two 10-foot (3-m) lengths
- Kite string
- Flashlight battery
- Carpenter's glue

Putting Electromagnets to Work

Our machine will use a double-coil magnet.

Bunch and glue 4 oatmeal boxes together so that they form a kind of square with a space in the center. Cut the piece of corrugated cardboard into a circle. At the center of the circle and at the center of the shoe box's bottom, trace the circumference of the cardboard tube. Use the craft knife to cut out this smaller circle.

Insert the tube into the space at the center of the oatmeal boxes. Drop the corrugated cardboard circle over the top of the tube so that it rests on the cardboard circle. Make sure the tube isn't higher than the shoe box. If it is, carefully trim it down to size with the craft knife.

Cut around the hole traced in cardboard so you'll have a cardboard washer to drop over the exposed tube. This ensures a smooth pivot of shoe box against gift box. Now you can place the shoe box over the tube, making sure it moves freely.

Electromagnetic Crane

Poke two sets of holes near the edges of the narrow mint box for the crane of our machine. Place the wooden spools between the holes, and insert pencils or dowels through holes and spools, making two pulleys. If you find that the dowel rolls inside the spool, use a rubber band for a spacer, twisting it around the shaft before reinserting the dowel into the spool. Finally, slide the grip tubes on all protruding dowel ends to keep them from slipping out of the box.

We'll modify the electric switch to make it more lightweight. Remove the shoe box lid, and place it horizontally before you. Near the right edge, insert two thumbtacks, spacing them no more than the paper clip's length. Carefully turn the cover over, and with pliers, bend the points of the thumbtacks to the side. Make sure you allow some play in the tacks, since you don't want them pressed too firmly against the lid. Turn the lid right side up, and slip a paper clip under one of the thumbtacks. Slide the paper clip to make sure that it makes contact with the other thumbtack.

With the shoe box lid still sitting horizontally before you, position the crane vertically in the exact center, the bottom of the narrow box flush with the lid's bottom edge. The crane's other side should stick out a few inches (7 to 10 cm) from the top edge of the lid. Glue the crane in place and allow it to dry. In the meantime, place the battery inside the shoe box on the right side, near the lid position of the switch.

Double-Coil Magnet

The machine uses a double-coil magnet. Wrap 50 turns of wire around the bolt, but this time leave 2 feet (60 cm) of wire at the ends. Do the same with the other wire. You should have a pair of wires, 2 feet (60 cm) long at opposite ends of the bolt.

Attach the first pair of wires to the contact of your switch and the other pair of wires directly to the battery (make a small hole in the shoe box's side to do this). Then, run a short pair of wires from the bottom of the second thumbtack to the battery. Place the lid (with crane) on the shoe box.

For the finishing touch, cut just enough kite string to keep your magnet ½ inch (1.25 cm) from the ground. Tie one end of the string to the top of the bolt, and tie the other end around the back spool. You're ready to show off your electromagnetic crane with the flip of a switch.

Homemade Galvanometer

You Will Need

For the Cabinet

- Two ¼-inch (.62-cm) pieces of plywood 5½ × 3¾ inches (13.75 × 9.3 cm) for sides
- ¼-inch (.62-cm) plywood 3½ × 3⁵⁄₁₆ inches (8.75 × 8.3 cm) for back
- ¼-inch (.62-cm) plywood 5½ × 3⁵⁄₁₆ inches (13.75 × 8.3 cm) for base
- ⅛-inch (.3-cm) plywood 3¹³⁄₁₆ × 1 inches (9.5 × 2.5 cm) for front
- ⅛-inch (.3-cm) plywood 3¹³⁄₁₆ × 2½ inches (9.5 × 6.25 cm) for top
- Two ⅛-inch (.3-cm) pieces of plywood 4 × ¼ inches (10 × 62 cm) for supports
- Wood strip 3 × ⅞ × ⅜ inches (7.5 × 2.2 × .95 cm) for scale support

For the Mechanism

- Clear acrylic 3¹³⁄₁₆ × 4⅛ inches (9.5 × 10.31 cm) for cover
- Miniature cabinet hinges with screws
- Small plastic or ceramic bead with screw-size hole
- White cardboard 3 × 1 inches (7.5 × 2.5 cm) for scale
- Roll of uninsulated copper wire or magnet wire
- 2 alligator clips
- Metal coat hanger
- Short fine nylon thread
- ½-inch (1.25-cm) diameter plastic-foam disk

Additional Materials

- Strong bar magnet
- Index card
- Six ½-inch round-head screws
- ¾-inch brads
- Short length of copper wire (24 gauge)
- ¼-inch stove bolt with nut
- 2 rubber washers
- Carpenter's glue
- Bottle cork about ½-inch (1.25-cm) diameter
- Drill with ¼-inch bit and needle bit for tiny holes
- Wire cutters

What It Does

A **galvanometer** is a sensitive measuring device used to detect small amounts of electric current. It helps you see electrical activity too small to light a bulb or drive a motor.

You can use a modified compass to measure small amounts of electric current, but building an accurate galvanometer isn't difficult. It adds a nice finishing touch when combined with other electricity projects.

Building a Galvanometer

Begin with the cabinet. Drill a pair of horizontal holes in the back panel 1 inch (2.5 cm) from the top and centered. Drill another hole in the top panel, centered, 1⅝ inches (4.08 cm) from the wide edge. Now join the base to the back panel using ½-inch brads and carpenter's glue. Add the two side panels and allow everything to dry before proceeding. We'll leave the top and front panels open for constructing the internal mechanism.

Finished Galvanometer

To construct the wire coil, cut the narrow side of the index card into ½-inch (1.25-cm) strips. Roll the card into a tube 1½ inches (3.75 cm) in diameter by 1¼ inches (3.12 cm) long. Glue the tube at the seam. When dry, wind about 150 turns of wire around the tube near the edge, leaving about 4 inches (10 cm) of wire at the ends.

Fan out the ½-inch (1.25-cm) strips and apply a little glue to the back of each strip. Attach the tube at the strip side inside the back panel of the cabinet about ¼ inch (.62 cm) from the base. Poke the wire ends through the hole drilled in the back panel, and attach alligator clips.

The magnetized plastic foam and indicator are the heart of the galvanometer. With a wire clipper, cut the metal coat hanger into ten 1-inch (2.5-cm) pieces. With a pole of the bar magnet, stroke five pieces in one direction and the other five in the opposite direction. The first five will have north poles at

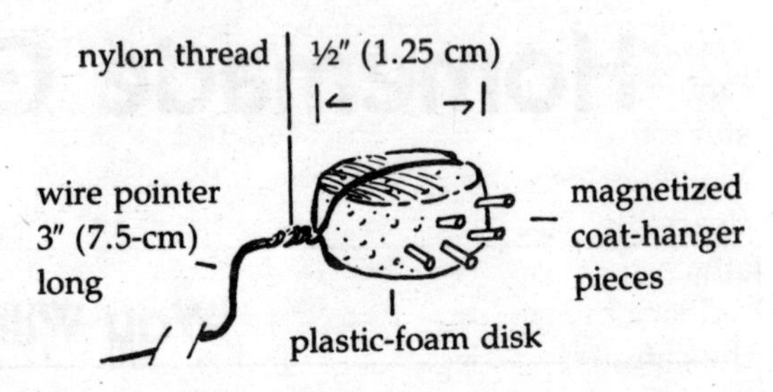

their points, and the second five will have south poles at their points. Insert the five north-pole magnets into the cork's left side and the five south-pole magnets into the cork's right side. Twist a piece of 24-gauge copper wire around the disk, back to front, so that you have a kind of pointer, 3 inches (7.5 cm) long. With the excess wire snipped off, make a small loop for tying the nylon thread.

For the support, bend a paper clip into a hook, and solder it into the slot of the ¼-inch stove bolt. Insert the bolt through a rubber washer, and slide it through the hole on the top panel. Put another rubber washer on the other end, and tighten the nut. You should be able to turn the stove bolt by loosening the nut a little—important for making adjustments in the magnetized disk later.

Tie the end of the nylon thread to the paper clip hook, and put the top panel in place. The magnetized cork and indicator wire should be perfectly balanced, and the disk should sit within the coil without touching the sides of the paper tube. Make necessary adjustments by bending the paper clip hook and wire loop. When satisfied, screw the cabinet top in place; do not use glue for the joints.

On the white cardboard, copy the scale.

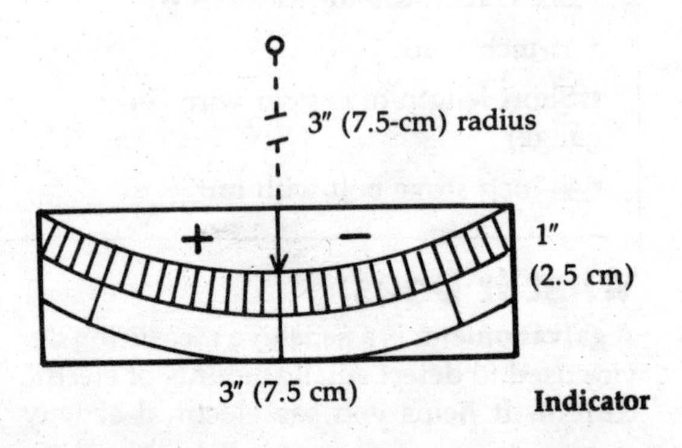

Indicator

Glue the wood strip to the base of the cabinet ⅛ inch (.31 cm) from the galvanometer's front, then glue the scale to the strip. Make sure the tip of the wire indicator just clears the cabinet. If everything checks out, attach the front panel to the cabinet.

Place the acrylic window over the front of the cabinet, and mark both hinge positions and placement of the bead for a handle. Place the window on a flat surface, and carefully drill small holes for the hinge screws. Do the same on the hinge side of the cabinet. Drill a larger hole for the bead, and attach the bead to the window with a stove bolt and nut to fit. Place the window back on the cabinet, and screw the hinges to the side panel to complete your galvanometer.

User's Guide to the Galvanometer

Since the galvanometer's magnetized disk acts as a compass, for accurate data you must orient the instrument so that its top is in a north–south direction. Use a compass at first, then mark the top for later reference. Adjust the indicator until it rests comfortably on the center line between the positive and negative quadrants.

When you connect a power source to the galvanometer, an electric current passes through the coil, turning it into an electromagnet. The stronger electromagnet overrides the magnetic field of the earth, and the indicator will swing to one side.

Connect a D-cell battery to the ends of wire sticking out of the back panel (remove the alligator clips and use a little tape). When the indicator points to the positive quadrant of the scale, mark the appropriate wire "positive" and the other wire "negative" to correspond to your battery terminals. This will come in handy later if you want to use this instrument to see which is the positive of an unmarked current supply.

Apply clear finish on the cabinet wood, if you wish. Your completed galvanometer will be useful in many projects.

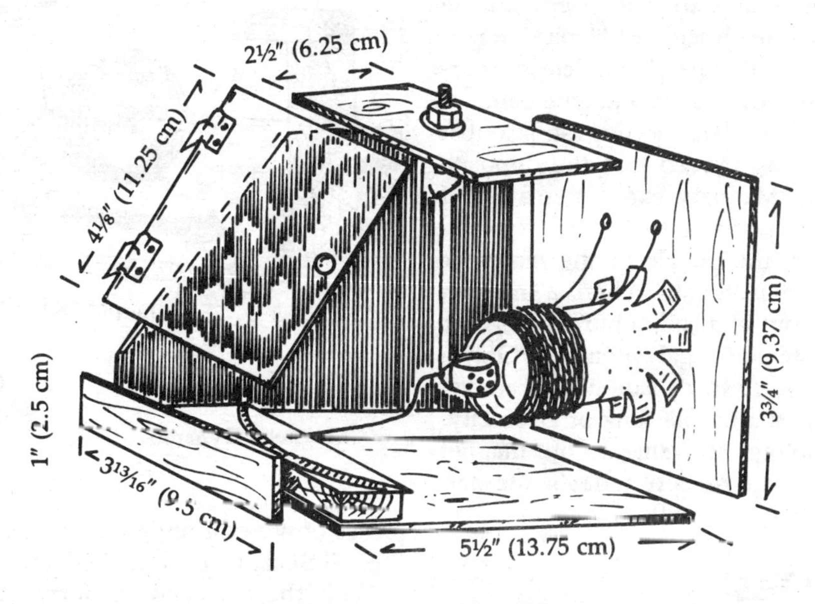

Galvanometer Assembly

Faraday's Discovery

You Will Need

- Two ½-inch (1.25-cm) plywood strips 2½ × 14 inches (6.25 × 35 cm)
- Medium nails
- Thin cardboard
- 2 rubber washers
- Pencil-size dowel
- Galvanometer (see Homemade Galvanometer)
- 10 feet (3 m) of uninsulated (enameled) copper wire, medium thickness
- Drinking glass
- Bar magnet
- ½-inch (1.25-cm) plywood platform 6 × 12 inches (15 × 30 cm)
- Two ½-inch (1.25-cm) plywood strips for platform feet
- Epoxy cement
- Drill
- Pliers

Alternating Current

Electromagnets owe their existence to the 19th century Danish scientist Hans Oersted, who discovered that passing an electric current through a wire creates magnetism.

Several years later, the English scientist Michael Faraday proved that the opposite was also true. Magnets, exposed to a coil of bundled wire, *create electricity*.

The key to the puzzle of the **magnetic dynamo** is movement. When a magnet passes in or out of a coil, a burst of electric current registers on a galvanometer. Only a rapid back-and-forth or side-to-side movement produces a usable flow of electricity. This alternating movement of magnet against coil is the basis of today's **alternating current** (AC).

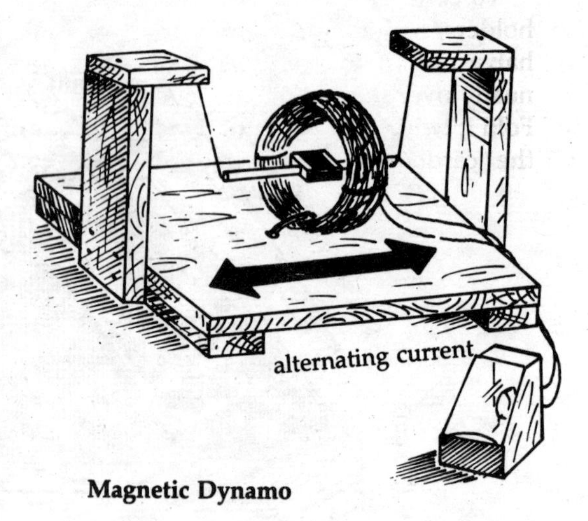

Magnetic Dynamo

Faraday Test

Wrap about 60 turns of wire around the drinking glass, leaving 1 foot (30 cm) of wire at the beginning and end. While you wrap, keep the wire close together to create a thick coil. Carefully slip the coil off the glass, and use the pliers to twist three or four smaller pieces of wire through and around it. Make the coil thick, firmly bunched, and compact.

With the plywood platform horizontal, position the coil vertically in the center. Draw a line on both sides of the coil to record its thickness. Drill a hole at the middle of each line. Fasten the coil to the platform with more wire, looped through the holes and tightened underneath.

Place the two ½-inch (1.25-cm) plywood strips underneath the platform and towards the edges to form feet. Drive nails through the platform to attach the feet.

Saw a 4-inch (10-cm) piece from the top of each 2½ × 14-inch (6.25 × 35-cm) plywood strip. Nail these smaller pieces to the ends of the larger pieces, perpendicularly, forming an L. Turn each L upside down, center it along the narrow edge of the plywood platform, and nail it in place.

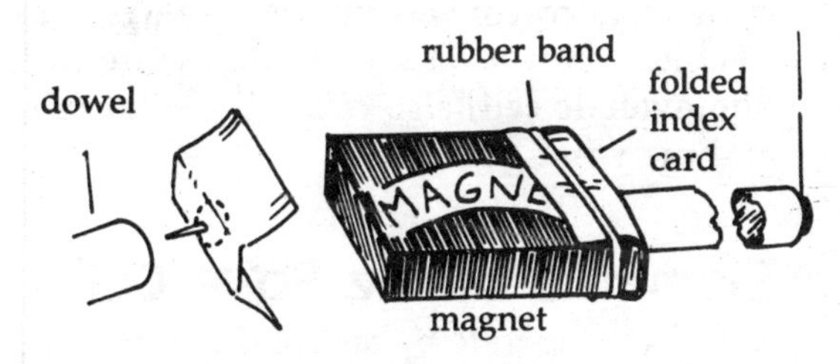

Pendulum Detail

To construct the pendulum and magnet holders, saw the dowel rod into two equal halves. Trim the cardboard so that it fits neatly over the bar magnet's narrow sides. Fold it twice, and push thumbtacks through the cardboard at the midpoint between folds. Where the tacks emerge, attach the dowel rods.

Secure the magnet by wrapping rubber bands or a strip of narrow masking tape around the cardboard. At the unattached end of each dowel rod, insert a thumbtack. Also attach thumbtacks to the undersides of the L pieces on the platform's sides.

Measure the distance between the underside of the L piece and the center of the coil. Cut two pieces of thread that length, tying one end to the dowel rod thumbtack and the other end to the L piece thumbtack. The bar magnet should sit on a kind of swing at the center of the coil. Connect two 1-foot (30-cm) lengths of unattached coil wire to the galvanometer.

In the Swing

For a demonstration of Faraday's discovery, slowly swing the pendulum back and forth horizontally through the coil. Notice how your galvanometer immediately registers current as the magnet passes through. Stop swinging and observe what happens. Start the pendulum swinging again, more rapidly this time, until the galvanometer indicates a steady flow of electricity. Imagine how excited Michael Faraday was!

Solar Cell

You Will Need

- Galvanometer *or* 0–100 microampere meter
- 6 feet (180 cm) of insulated copper wire—3 feet (90 cm) of blue and 3 feet (90 cm) of red
- Thin-ply aluminum foil
- Portable radio (discarded)
- Electrical tape
- Lacquer thinner
- Cardboard strip
- Soft rag

Natural Power

Since electrical energy, based on burning fossil fuels once considered inexhaustible, pollutes our environment and destroys valuable natural resources, scientists have sought alternative, clean natural sources of inexhaustible energy. The answer was right in their backyards.

Energy produced from natural sources has been around hundreds of years—the simple water wheel or windmill, for example. With the discovery of electricity and the invention of the electrical generator, **hydroelectric power**—or electricity produced from the tremendous force of water across dams—was widely used. More sophisticated technologies and the discovery of new materials soon led to more direct and efficient ways to convert nature's energy into electrical power. Two examples are the **photovoltaic cell,** also called the **solar cell,** and the **wind turbine.**

Elements of the Solar Cell

Solar cells operate by taking the sun's radiant energy and converting it directly into electric current. The element **selenium,** a Greek word meaning "moon," was the first substance from which solar cells were made. About 20 years ago, camera light meters were powered by selenium cells. Since selenium only converts a small percentage of radiant energy into electric current, scientists looked for more efficient materials. The element **silicon,** one of the most abundant elements on earth, converts a much higher percent of the sun's rays into electricity. That's what we use today in solar cells.

A solar cell consists of a very thin wafer of silicon over which a delicate metal grid is

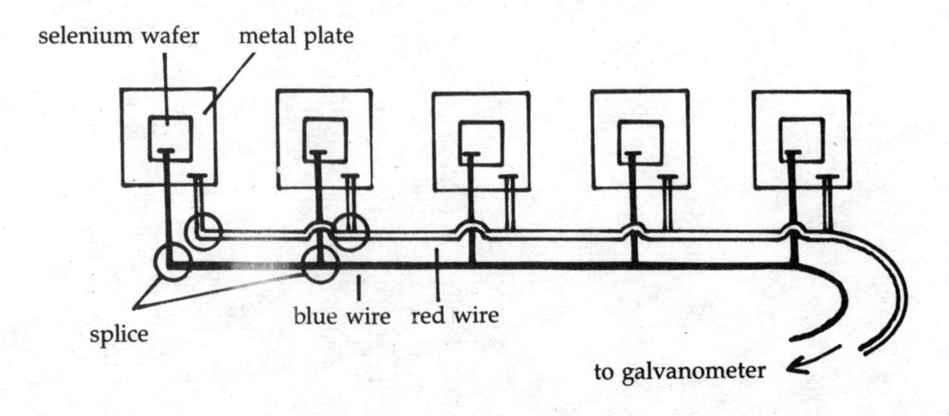

Solar Cells Connected in a Series

applied. The grid harvests electrons from the silicon without shading it from the sun. One cell by itself produces too little power for practical use, so many cells are usually connected in a series.

Constructing a Solar Cell

Although you can buy solar cells in many electronics stores, a discarded older model portable radio can provide the raw materials for constructing your own solar cell series. The current produced will be very small, but you'll get a response from your galvanometer.

Open the radio's back with a screwdriver and remove the selenium rectifier from inside (be certain to unplug the radio or remove batteries first). The rectifier has a panel of six plates, each with a copper brown selenium wafer against a metal base, soldered at the ends. With a little cleaning and the right connections, each plate can function as a cell.

Carefully cut the plates from the rectifier, and remove any paint with lacquer thinner on a soft cloth. Avoid rubbing too vigorously—you don't want to scratch the delicate selenium surface. Glue the plates, metal side down, to the strip of cardboard. Space them evenly.

Cut the red wire into six 2-inch (5-cm) pieces and the blue wire into six 4-inch (10-cm) pieces. Lay the pieces out in pairs. Cut the thin foil into small squares, just large enough to act as contact pads for the wire tips. Now, with a tiny bit of solder, connect one end of both red and blue wires to the contact pads, leaving the other ends free.

With small strips of tape, attach the contact pads of the blue wires to the selenium squares (negative). Tape the contact pads of the red wires to the metal bases (positive). Splice the six blue wires to a longer piece of blue wire, functioning as the negative lead wire for the galvanometer. Splice the six red wires into a larger piece of red wire, functioning as a positive lead wire.

Shine a strong light on your panels. Watch the needle swing as it registers an electric current. With a strong light source and many more cells connected in series, you could even light a 6-volt flashlight bulb or power a small motor.

Selenium rectifiers may be difficult to find. They're in radio models built before 1970, but you may have to search in stores that sell used electronics parts. If you want to design a project that utilizes a great many cells, buy silicon cells from a science supply house. Most come already wired and easily connect in series.

Wind Turbine

You Will Need

- Galvanometer
- Model airplane propeller about 6 inches (15 cm) long
- 2 nails 1 inch (2.5 cm) long
- 2 nails 3 inches (7.5 cm) long
- 4 small nails or brads
- Small bar magnet, 1 inch (2.5 cm) long
- 2 brass strips, 1½ × 4 inches (3.75 × 10 cm)
- 10 feet (300 cm) of uninsulated (enameled) copper wire
- Germanium diode
- Piece of plywood or wood block, 3½ × 5 inches (8.75 × 12.5 cm)
- Drill
- Electrical tape
- Carpenter's glue

Modern Windmills

An elegant modern application of the Faraday principle is the wind turbine. Wind turbines—really modern windmills—rise like sunflowers along highways and provide inexpensive energy. More and more wind turbines have begun to appear as costs decrease and technology improves.

Each windmill is an **electric generator,** since movement of a powerful magnet attached to the propeller creates electrical energy in a coil. A small electronic apparatus, the **diode,** keeps the current flowing in a single direction.

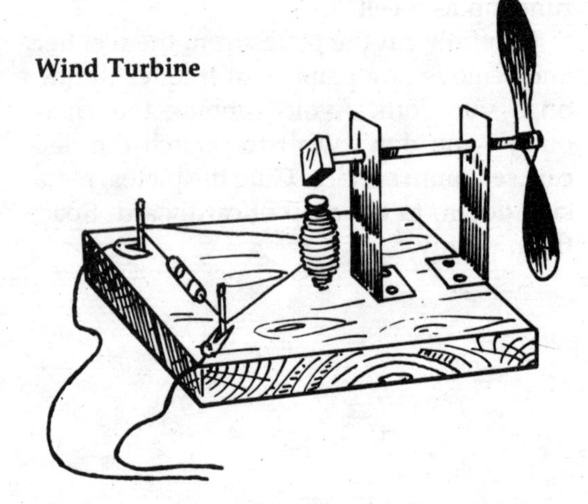

Wind Turbine

Constructing a Wind Turbine

Wrap about 200 turns of wire around one of the 3-inch (7.5-cm) nails for the coil. Leave 1 inch (2.5 cm) of space at the nail's bottom to hammer it into the wood base. Also, remember to leave a few inches (several cm) of free wire at the coil's beginning and end for connections. Twist the wires' ends together once or twice to prevent unravelling.

Hammer the large nail to the center of the wooden base, and hammer two smaller nails behind it so that the three nails form a triangle. Loop the free ends of coil wire once around each smaller nail (soldering, if necessary) and connect them with the diode at the base of this triangle of nails.

With carpenter's glue, attach the bar magnet to the head of the other 3-inch (7.5-cm) nail. Allow the glue to dry thoroughly. Center the magnet on the nail head so that it spins cleanly when the nail—now a propeller shaft—revolves.

For the supports, measure ¾ inch (1.8 cm) from the bottoms of the brass strips, and bend each strip 90 degrees at that place. Hold one of the strips, with bent side down, against the wooden base to determine where to drill holes for the propeller shaft. Make sure the end of the shaft with the attached magnet just clears the coil's top. If your magnet is a little longer than 1 inch (2.5 cm), you may need longer strips for supports or need to build up existing supports with thin sheets of masonite or linoleum.

Carefully drill holes in both supports, making sure the holes line up. Then screw the supports to the wooden base directly in front of the coil, with bent ends facing inward. Poke the shaft through the holes so that the magnet sits directly above the top of the coil. Wrap a little electrical tape around the shaft on both sides of each hole so that both shaft and magnet stay in position.

Finally, attach the propeller to the end of the shaft. There should be a little hole in the back of the propeller; if it slips, use tape for a spacer.

Windy Results

To test your wind turbine, connect the wires of the galvanometer to the small nails. Position an electric fan in front of the propeller and switch it on. Watch the compass needle jump as your turbine creates electricity from the wind.

Light & Sound Show

Bending Beams of Light
Shimmering Soap Bubbles
Giant Soap Sheet
Peeking Periscope
Pinhole Camera
Tiny Kaleidoscope
Polarized Light Box
Dancing Lights
Singing Glasses
Bottle & Pipe Trombone
Sound Waves from a Tuning Fork
Echoing Hose Phone
Boom-Box Tube
Edison's Reproducer
Whispering Balloons
Parabolic Sound-Collecting Dish

Bending Beams of Light

You Will Need

- Set of 4 lenses
- Small square aquarium
- 1 sheet of black poster board
- Craft knife
- Pocket comb
- Electrical tape
- Hooded lamp
- Copper wire (24 gauge)
- Small nail or brad
- Black tempera paint
- 2 cups of whole milk

Refraction

One of light's many unusual properties is that it can be bent, or **refracted,** when it passes through various substances. See, for example, how the bottom of a pencil placed on a glass of water does not seem connected to the top. Light travels *faster* in air than in water, so it bends slightly when passing from one medium to the other. You can demonstrate how light bends as it passes through four basic lens shapes. After experimenting a little, you should be able to draw conclusions from your results.

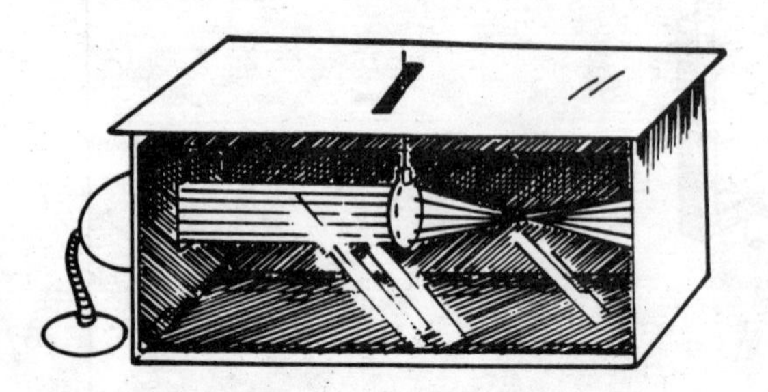

Light Box

Constructing a Light Box

Paint the outside of an aquarium's sides and back black. After the paint dries, use a craft knife to scrape a vertical window on one side of the aquarium, about the size of a pocket comb. With electrical tape, attach the comb to the glass so that the teeth completely cover the window.

Cut the black poster board so that it fits over the aquarium's top. Use the craft knife to cut a vertical slot in the center of the board, just wide enough for the lenses to slide through. Fill the aquarium with water, and add two cups of whole milk to make the water slightly cloudy.

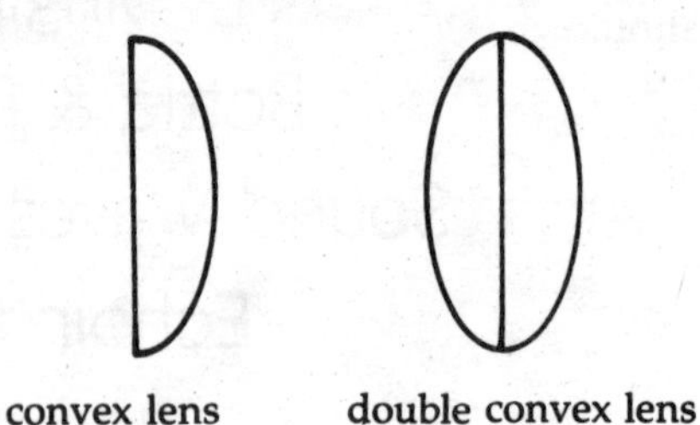

convex lens　　double convex lens

Positive Lenses

Examine your lenses. You need a plain convex lens, a double convex lens, a plain concave lens, and a double concave lens. **Convex** and **double convex lenses** are called *positive* because beams of light come together when they pass through. **Concave** and **double concave lenses** are called *negative* because beams of light expand when they pass through.

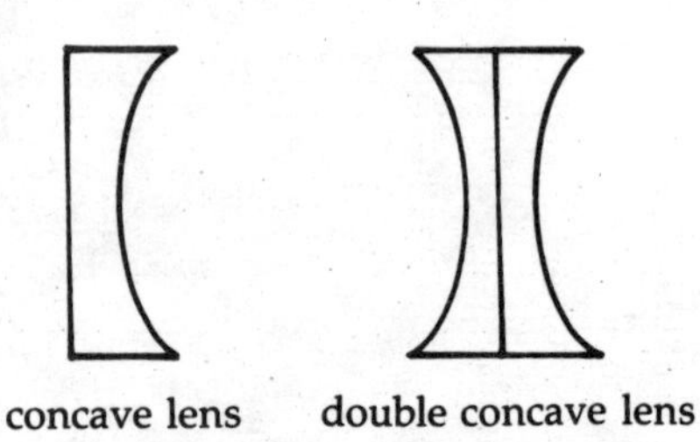

concave lens　　double concave lens

Negative Lenses

Mounting the Lenses

To prepare your lenses for the light box, you have to mount them on a wire frame. Carefully twist some 24-gauge copper wire around the edge of each lens. To make sprockets, cut small pieces of wire, slide them under the twisted wire, and double them over. Attach a longer piece of wire to the mounted lens as a sort of handle.

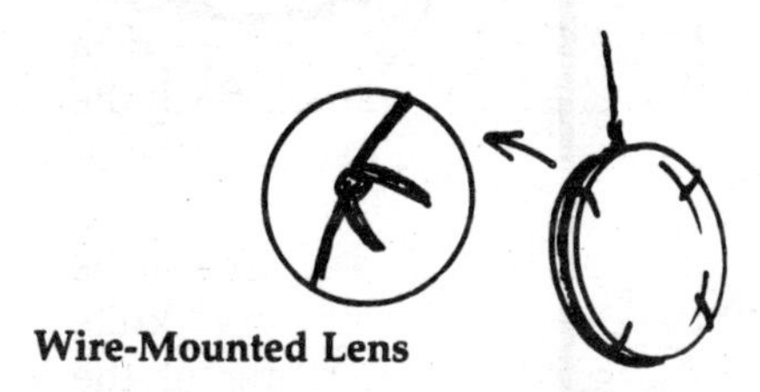

Wire-Mounted Lens

Setting Up

Place the hooded lamp against the pocket comb. Light coming through the comb's teeth should make parallel beams through the cloudy water. These light beams are the result of the **Tyndall effect:** tiny protein and fat particles in the milk reflect the light as it travels through the liquid.

Carefully slide one lens through the slot at the top of the aquarium and position it in front of the light beams. Observe the light beams as they pass through the lens and out again. Notice whether the beams expand or come together. Where the beams come together is the **focal point** of the lens. Though the stripes of light may not reveal it, every lens has a focal point. With negative lenses, you find the focal point by tracing the expanding stripes back through the lens to the point where they come together.

Diagram each beam pattern as you insert the different lenses and indentify each focal point. Allow viewers to draw their own conclusions about lens shapes and paths of light.

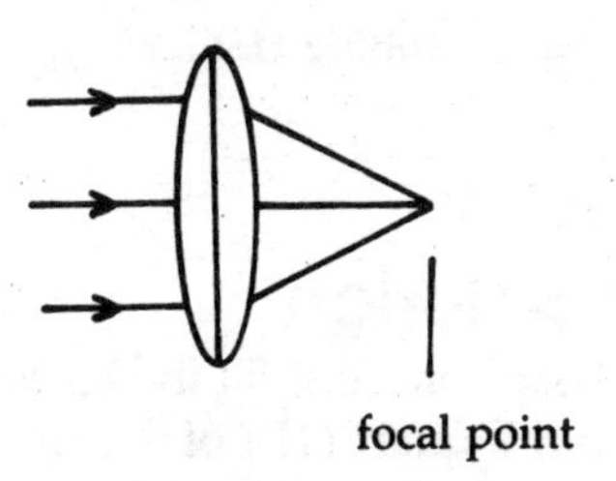

Double-Convex Lens

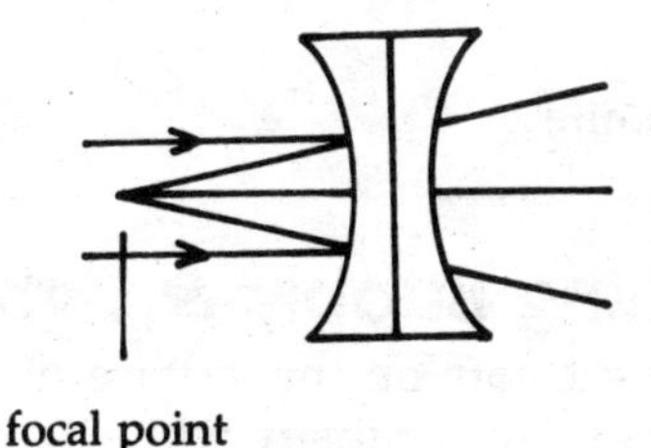

Double-Concave Lens

Shimmering Soap Bubbles

You Will Need

- Soap solution made from dishwashing liquid
- 2 to 3 tablespoons of sugar or corn syrup
- Bucket or wide bowl
- Drinking straw
- Flexible wire in 2-foot (60-cm) lengths
- Plywood or wood block 5 × 5 inches (12.5 × 12.5 cm)
- Drill
- Wire clippers
- Pliers
- Craft knife

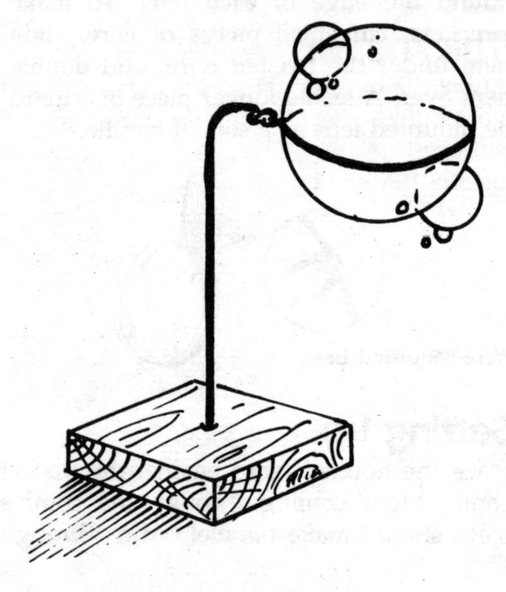

Bubble Holder

Something Bubbly & Light

The swirling colors on the surface of soap bubbles come from ordinary white light, reflected from opposite sides of the bubble's surface. But something fascinating happens to the light before it reaches our eyes.

Bubble Holder

A bubble holder will allow us to enjoy those colors as they shift and change. Drill a hole in the center of the wood base just large enough to fit the wire snugly. Make the bubble holder by taking one end of the 2-foot (60-cm) wire and forming it into a loop (wrap the wire around a jar or can to shape it). Using pliers, close the loop by twisting the free end against the straight wire. Measure about 2 inches (5 cm) along the straight wire from the twist, and bend the remaining wire 90 degrees. Insert the wire's end into the hole in the wood base, making sure the holder stands securely.

Soap Solution

For the soap solution, fill the bucket or bowl with about a quart (1L) of warm water and add dishwashing liquid until the water feels slightly slippery. Finally, add 2 to 3 tablespoons (30 to 45 ml) of sugar. The sugar makes the solution thicker and adds to the life of your soap bubbles.

Bubble Pipe

To make an excellent soap pipe, take the drinking straw and carefully make four ½-inch (1.25-cm) slits at one end with the craft knife. Bend the sections out into a kind of flower. Dip the flower end into the soap solution and gently blow through the opposite, mouth end, until you have a good size bubble.

While stopping the mouth end with your finger to keep the bubble inflated, carefully move your bubble to the wire loop, and drop it down so that the bubble's sides stick to the loop's sides. Gently twist the pipe against the wire, and the bubble will disconnect and remain in the loop.

Bubble Shapes & Thicknesses

Enjoy the brilliant colors as they swirl over your bubble. Each color indicates a different surface thickness. Yellow, for instance, indicates that the bubble is 1⁄30,000-inch (1⁄12,000-cm) thick in that area—or 1,000 times thinner than a piece of paper.

With the remaining pieces of wire, create different shapes to see how soap film sticks to each one. Three-dimensional shapes such as cubes, pyramids, even squiggles will be connected every which way by sheets of soapy film.

For your display you could have several wire shapes and buckets. That way, viewers can participate and enjoy the soap bubble colors.

Giant Soap Sheet

You Will Need

- 3 feet (90 cm) of ½-inch (1.25-cm) diameter plastic PVC pipe
- Plastic planter trough 3½ feet × 6 inches (105 × 15 cm)
- Small clothesline pulley
- 2 small metal springs
- Thin nylon cord 10 feet (300 cm) long
- 2 eye screws
- Tire repair kit—rubber and cement
- Table vise
- Plywood platform same width as planter box, but at least 6 inches (15 cm) longer
- 16 feet (480 cm) of 1 × 4 wood for frame, cut into two 4-foot (120-cm) sections, four 1-foot (30-cm) sections, and one section as long as the plywood platform
- Kite string or thin cord
- Small fisherman's sinker weight
- Dishwashing soap-and-water solution
- Drill
- Craft knife
- Saw and mitre box
- Nails
- Carpenter's glue

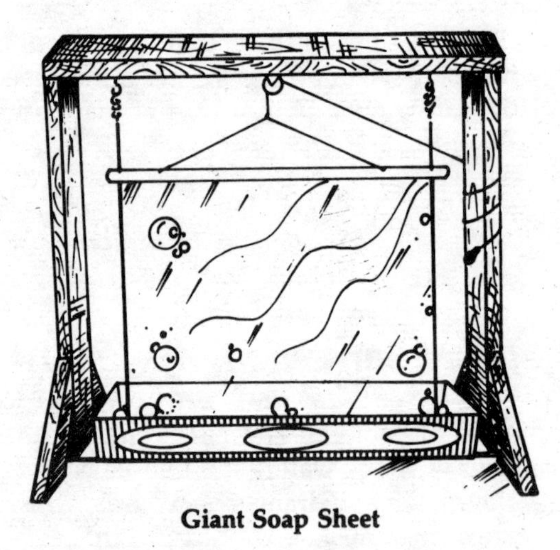

Giant Soap Sheet

Soapy Film

A soapy film, stretched into a large sheet, allows you to examine closely the colors and patterns of reflecting light. With the help of a few drinking straws, the sheet also provides a launching pad for giant bubbles. The sheet is made by dipping the long PVC pipe into the trough of soapy solution, then slowly drawing it out. The pipe is hung and carefully balanced with kite cord.

Construction

Measure ½ inch (1.25 cm) from the pipe's ends and drill holes there. Drill a pair of holes through the pipe, taking care to align them so that the kite string can slip through easily. Use a vise to secure the pipe while you drill.

Measuring again from the ends, use a craft knife or saw to make 1-inch (2.5-cm) slits along the pipe's top across the top holes.

The slits should be just wide enough to pinch and secure the kite string. Pull the string through the pipe, then jerk it up through the slits, beyond the holes. Loop the string around and tie it according to the diagram.

A simple slip knot allows you to adjust the angle at which the pipe hangs. Keep the remaining string attached to the roll for now.

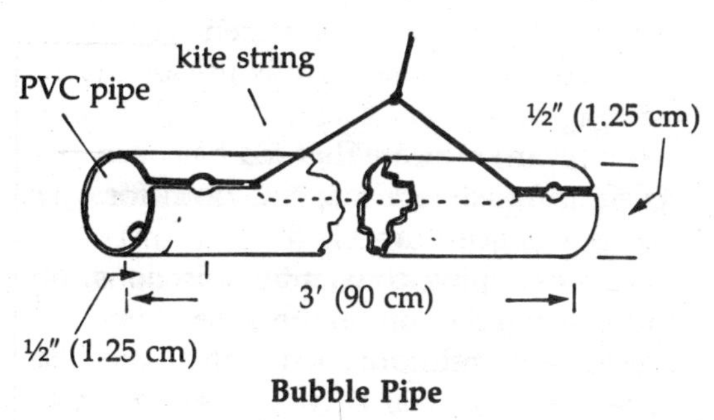

Bubble Pipe

Determine the exact center of the trough by dividing and marking the bottom surface into halves, horizontally and vertically. Make corresponding measurements on the piece of wood used for the top section of the frame. After finding and marking the middle of the pipe, align the pipe horizontally in the trough so that the holes face up. Push a pencil through the holes, marking their position on the trough. Align the pipe with the wood and repeat the procedure. Attach eye screws at the hole marks on the wood, and attach the pulley at the center mark.

Assemble the rest of the frame according to the diagram, using carpenter's glue or small nails to attach the mitre-cut support pieces.

Now we'll construct runners for the pipe by securing nylon cords to the bottom of the trough and stretching them up through the frame. Tie the end of the first cord to a small piece of stiff wire. A piece of paper clip works fine. From the tire repair kit, cut out a square of rubber large enough to completely cover the piece of wire. Make a pinhole in the center of the square, and pass the cord through the hole until the wire sits against the rubber. Apply the tire cement to the wire side of the square, and press the square against the trough's bottom, centering it so that the cord sits over the hole mark.

Repeat the procedure for the second cord, allowing the glue to dry. Depending on the plastic used for the trough, the tire cement will make a secure or weak bond. If the rubber square separates from the plastic, substitute epoxy glue for tire cement.

Place the trough on the plywood platform. Thread the unattached ends of nylon cord through the pairs of holes at opposite ends of the pipe. Finally, tie the end of each cord to a spring. Hook the springs to the eye screws at the frame's top. The trough should now hang like a swing from the frame, clearing the platform by less than ½ inch (1.25 cm).

With the pipe lying at the trough's bottom, measure the length of kite string required to pass up through the pulley and around it. Cut the string from the roll at that point, and tie a small sinker weight to the end.

Test the pipe for proper weight and balance. Place a flat object in the trough, something heavy enough to push it down against the platform, making the runners taut. Carefully pull the weighted end of the pulley string, raising and lowering the pipe. If the movement is smooth, the pipe is weighted properly. If not, insert additional sinker weights (or pennies) into the pipe's ends.

Remove the object and fill the trough with soapy solution. Slowly raise the pipe. How can you help but marvel at that beautiful, shimmering sheet that forms between the pipe, runners, and solution? Wrap the weighted end of the string around the frame to keep the pipe suspended. Examine the film carefully.

Soapy Sandwich

What you see is a soapy sandwich—two layers of soap molecules forming a skin around a layer of soapy water. The middle layer of water moves and drains downwards, which creates a transparent medium of ever-varying thicknesses, not unlike a flexible lens.

When light strikes the soapy front layer, some of it reflects back to your eyes immediately while the rest continues to the soapy rear layer before bouncing back through the water layer. The light reflected from both the outer and inner layers combines and reaches our eyes at the same time. But during that short trip, something amazing happens.

Light-Wave Interference

As the light waves combine, they either reinforce or cancel each other out. We know that waves of light, like waves of sound, consist of crests and troughs. When two waveforms line up "in phase," their crests and troughs together, we call this **constructive interference.** The color created by one waveform is reinforced by the other waveform. When two waveforms are "out of phase," we call this **destructive interference.** The color created by one waveform cancels out the color of the other waveform.

All colors of the spectrum derive from the primary colors red, yellow, and blue. If the thickness of the soapy film is just right to cause the destructive interference of one of these colors, you see a mixture of the remaining colors. For example, swirls of blue green (cyan) indicate that red has been cancelled, purplish red (magenta) indicates that yellow has been cancelled.

Viewers enjoy constantly shifting colors and graceful motion on the soapy film. This process of reinforcement and cancellation—a kind of war between light waves—occurs millions of times per second.

Peeking Periscope

You Will Need

- Stiff cardboard or poster board 11 × 14 inches (27.5 × 35 cm)
- 2 square or rectangular pocket mirrors
- Craft knife
- Carpenter's glue

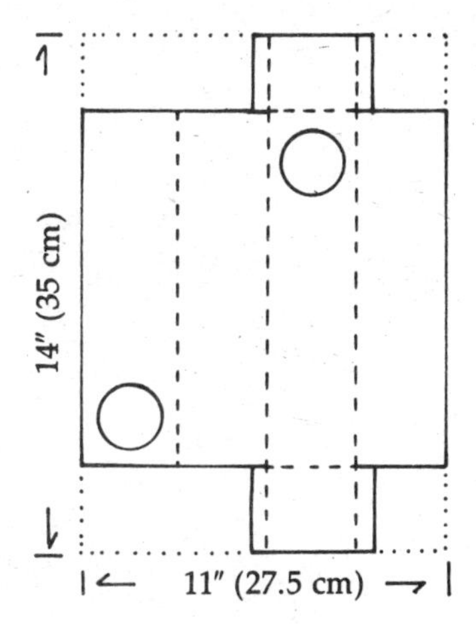

Periscope Pattern

A Matter of Reflection

Mirrors allow us to stretch our senses beyond their normal limits. From a giant telescope to a submarine's **periscope,** mirror devices utilize essential properties of light and provide valuable information. Physicians use an advanced form of the periscope to look inside the human body—a flexible plastic (**fiber optic**) tube that reflects light through it.

Trick with Mirrors

A basic periscope consists of a hollow shaft with mirrors positioned at opposite sides. This design will take about an hour to complete.

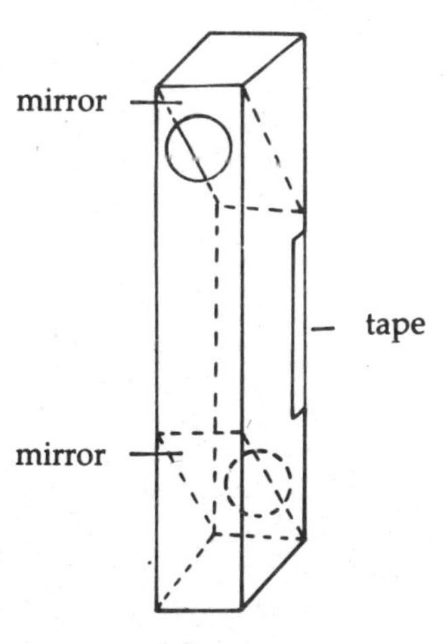

Completed Periscope

With ruler and pencil, copy the periscope pattern on the cardboard, including dotted lines. Then, carefully cut it out.

The dotted lines indicate places to fold the cardboard. To make a clean fold, first score (make a shallow slice) with your craft knife along the dotted lines. Avoid cutting through the cardboard. Flip the cardboard over and carefully bend it away from you; the pattern should begin to fold along the scores. Then, with a ruler's edge, flatten the cardboard along the folds.

Close the box along the folds, leaving the top and bottom flaps open. Stretch a long piece of masking tape where the edges join. Then, place the box on its back with the top hole on your right side.

Attach a strip of tape to the edge of one of the mirrors. If the mirror is rectangular rather than square, apply the tape to the narrow edge. Carefully slide the mirror into the box's end with its reflecting side up, until you see only mirror when looking through hole. Now press the tape down to anchor the mirror in place.

Turn the box end up, with the mirror at the top. Gently push the mirror so that it falls forward at a 45-degree angle towards the hole. Make sure you can close the box lid over the mirror. If the mirror sticks out too far, tape it farther down in the box.

Repeat this procedure for the second mirror on the box's opposite side. If everything fits snugly, apply masking tape along all open edges to keep light from leaking into the box.

Peer Advantage

Invite your viewers to try out the periscope. It can be used to peer, not only above, but around obstacles as well. You could also suggest some applications of the basic periscope design.

Pinhole Camera

You Will Need

- Round oatmeal box (18-ounce or .5L size)
- Thick shirt cardboard
- Black tempera paint
- Small paintbrush
- Aluminum foil
- Craft knife
- Carpenter's glue
- Light-sensitive paper

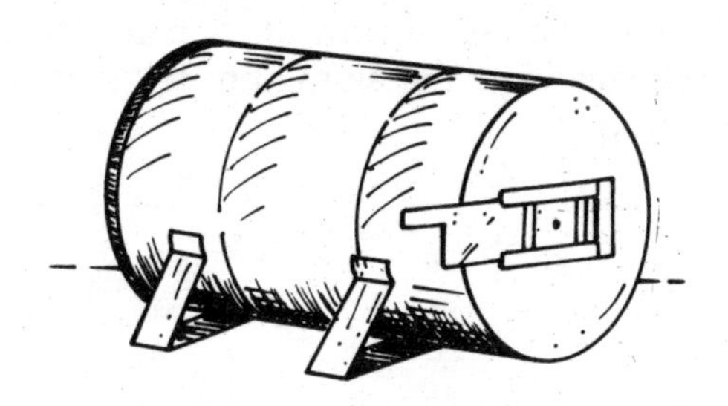

Pinhole Camera

Lights, Camera

A simple camera demonstrates basic principles of light. This model will provide reasonably good pictures that you can display beside it.

. . . Action

Remove the plastic lid from the oatmeal box (cylinder) and paint it black. Then paint the interior of the cylinder black. Allow everything to dry overnight.

In the center of the cylinder's bottom, cut a 1 × 1-inch (2.5 × 2.5-cm) square opening with a craft knife. Cut a piece of aluminum foil slightly larger than the square, and secure it over the opening with tape along the edges. Try not to wrinkle the foil too much as you tape it, and make sure the tape lies flat against the cardboard. Finally, make a small pinhole for the lens in the center of the aluminum foil.

Use cardboard for the lens shutter and legs to steady the camera.

Shutter to See Results

The shutter consists of a sliding panel that fits between two tracks over the lens. Draw the shutter patterns on cardboard and carefully cut them out. With a small amount of glue, attach the narrow strips along the edges of the wider strips. Make sure that no glue bleeds from under the smaller strips.

Position the shutter piece over the camera lens so that it covers the lens completely and the narrow tab sticks out from the camera's side about ½ inch (1.25 cm). Hold the shutter securely in place, and trace around it with pencil. Glue the strips of cardboard around the shutter's outline, again checking for excess glue. Allow everything to dry. Slide the shutter, flat side first, into the open

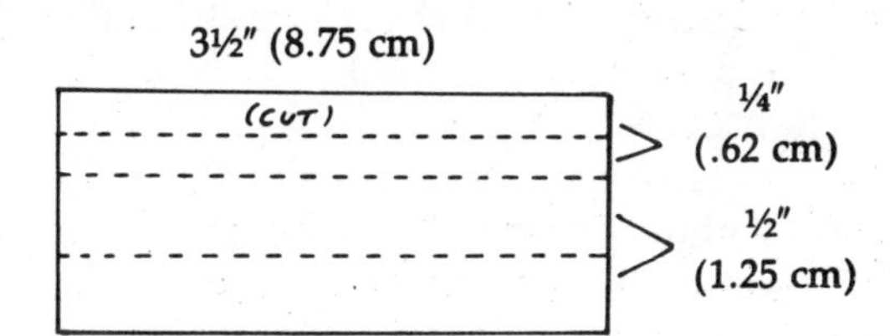

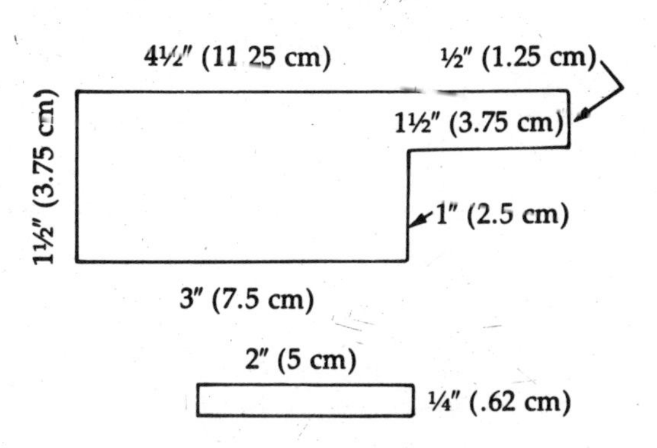

Template for Sliding Shutter

end of the track. Make sure it slides freely and sits tightly against the foil.

If everything seems to work, carefully coat the outside of the camera, including the shutter and track, with black paint. Make sure you leave the aluminum foil lens unpainted.

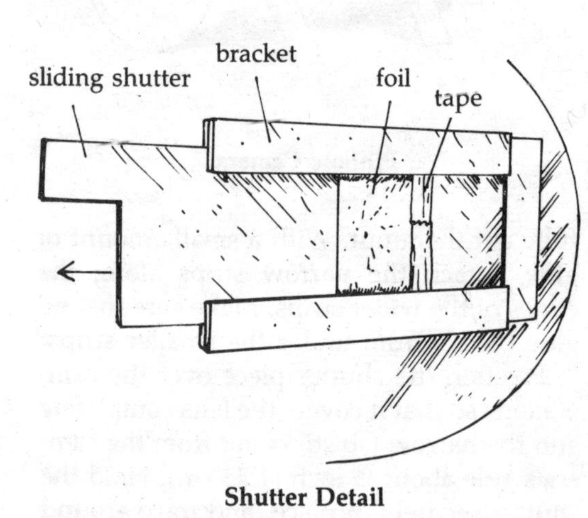

Shutter Detail

Legs to Stand On

Now for the legs. Cut two 1 × 10-inch (2.5 × 25-cm) strips from the cardboard. On each strip, measure 2 inches (5 cm) from the ends and draw a line. With a craft knife, carefully score (make a shallow cut) along those lines so that you can bend the cardboard without difficulty. Flip the strips over and measure ¼ inch (.62 cm) from the ends, folding as before. Fold the cardboard on the opposite side of the score. Paint the strips on both sides and allow them to dry.

Attach the folded strips to the body of the camera with glue. The glue should set quickly enough to allow your camera to stand on its legs undisturbed while you prepare the photosensitive paper.

Darkroom Prep

The photosensitive paper will be the film of our camera. In a darkened room, cut the paper down into 3 × 3-inch (7.5 × 7.5-cm) squares—just large enough to fit inside the camera. Bring the camera into this darkened area and remove the plastic lid. With a small bit of tape, make a tape loop, and press it against the inside of the lid. Press a square of "film" against this tape, shiny side out, and close the lid. Make sure the shutter is closed, and bring your loaded camera out into the daylight.

Light & Dark Subjects

Place the camera on a flat surface, and point it towards something you want to photograph. Open the shutter by pulling the tab, and wait about five seconds. Then close the shutter and move your camera inside.

In a dark place, remove the film and wrap it in several sheets of aluminum foil to keep out the light. The chemicals, containers, and instructions for developing film can be found at your local photo supply store. Or take your exposed film to a photo supply store, and have professionals develop it for you.

Tiny Kaleidoscope

You Will Need

- 3 glass microscope slides
- Black paint (not water-based)
- Brush
- Newspaper
- Black electrical or masking tape
- Bits of colored cellophane
- Clear contact paper (laminated sheets)

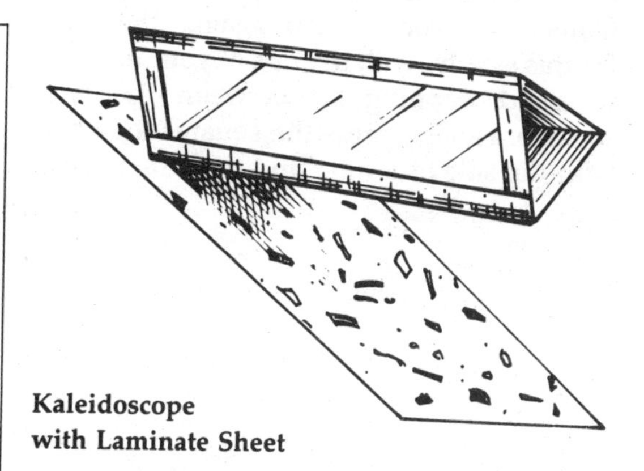

Kaleidoscope with Laminate Sheet

Special Optical Effects

Mirrors not only provide information, they can produce beautiful optical effects. When scientists learned how to grind mirrors into high-quality reflectors, a simple toy—the **kaleidoscope**—was invented.

Making the Prism

To avoid the difficulty of cutting mirrored glass, use microscope slides for reflecting surfaces. They work just as well. Lay the three slides side by side on newspaper, and apply a coat of black paint to one side only. Spread the paint thickly enough to avoid streaking (you may need several coats). Allow the slides to dry overnight.

Touch the paint for tackiness before you proceed. If thoroughly dry, put the slides together lengthwise in the shape of a prism, painted sides out (see illustration). Loop a rubber band tightly around the prism to hold it together.

Stretch a small piece of tape around the prism to one side of the rubber band. Then, remove the rubber band and stretch two more pieces of tape, each around the ends of the prism. Continue securing your kaleidoscope prism with strips of tape on the joints between the slides. Measure the tape carefully, and trim away any excess tape so that nothing obscures the open ends.

Simple Test

To test your kaleidoscope, close one eye and look through an open end with the other eye. Whatever passes through your field of view is multiplied nine times to form wonderful symmetrical patterns.

Translucent Bits & Pieces

Collect as many small, interesting objects as you can, keeping in mind that they must be translucent—allow light to pass through them. Bits of colored cellophane from candy wrappers, small costume jewelry stones, tiny leaf and flower parts make interesting kaleidoscope patterns.

Cut the contact paper into 3 × 3-inch (7.5 × 7.5-cm) squares. Peel the backing from one of the squares and carefully stretch it out, with the gummed side facing you, against the newspaper. Use a little tape on the corners to keep the sheet from curling up.

Now, take a pinch of your colored objects and sprinkle them across the contact paper. Try to get an even, thin coating of objects, but avoid bunching too many things together since you want light to show through.

Next, peel the backing from a second square of contact paper and hold it over the coated square, sticky side down. Carefully press one edge of the second square against one edge of the first square; then, slowly flatten the second square against the first. Do this slowly, with the tips of your fingers, to avoid trapping air between the two squares. Finally, press the squares together with the side of a ruler and trim the edges.

Operation: Kaleidoscope

Hold the square against one end of the kaleidoscope, and peer through the other end. Move the square slowly around, and enjoy watching the amazing shapes and patterns.

Look at newsprint, fabric, or wood grain through the kaleidoscope. If you have adequate lighting, you can look at many materials. Provide an assortment of objects for viewers.

Polarized Light Box

You Will Need

- 3 wooden 12 × 16-inch (30 × 40-cm) picture frames with glass or plexiglass covers
- 12 × 16-inch (30 × 40-cm) plastic sheet polarizing filter
- 5 × 5-inch (12.5 × 12.5-cm) plastic sheet polarizing filter
- 6-inch (15-cm), ½-inch (1.25-cm) diameter wooden dowel
- Small round-head screw and washer, less than ½-inch (1.25-cm) diameter
- Clear cellophane wrap
- Cardboard box, large enough for 11⅞ × 15⅞-inch (29.7 × 39.7-cm) hole
- Sheet of ¼-inch (.62-cm) thick balsa wood, cut into strips
- Craft knife
- Rubber cement
- Carpenter's glue
- Duct tape
- Black and white tempera paint
- Small desk lamp, preferably with hooded shade

Polarized Filter

White light is a chaotic combination of color **wavelengths** which you can separate with the aid of a prism. Scientists have discovered other ways to separate light by creating materials with light filtering—or **polarizing**—properties. **Polarizing filters** allow certain wavelengths of light to pass through while keeping others out, and they make possible such diverse and useful things as liquid crystal displays of digital watches, safe airplane windshields, and glare-proof sunglasses.

Think of a polarized filter as a window shutter. In the most common variety, long chains of iodine molecules remain fixed in place, allowing different wavelengths of light to pass through according to the filter's position. In your wristwatch's liquid crystal display, a weak electric current causes the suspended crystalline molecules to turn in the same direction, closing the shutter and blocking out light completely.

Polarized filters have many useful applications, but we'll construct a device that simply demonstrates the beauty of polarized light.

Polarized Light Box

Making a Glass Sandwich

Begin by removing the glass from the 12 × 16-inch (30 × 40-cm) frames and setting it aside. Be very careful when handling sheets of glass: *Wear gloves and pick each one up horizontally, holding it along opposite edges. The edges can be sharp.*

Discard all but one frame. Using a pencil or crayon, draw a design on white paper (12 × 16 inches or 30 × 40 cm), that's not

Draw design on cardboard.

too intricate or complicated to cut out of the cellophane. Then, place a sheet of glass over the picture, and keep rubber cement, cellophane wrap, and scissors close by.

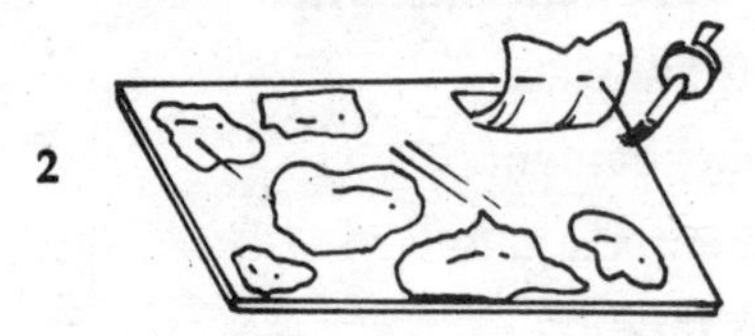

Glue pieces of cellophane to glass, following design.

Now, carefully cut the cellophane wrap into shapes that you can press onto the glass (see illustration). Use a little rubber cement at the edges of each shape to prevent the cellophane from slipping off the glass. Allow yourself some freedom in following your design. The shapes do not have to be perfectly accurate, and the more you layer shape upon shape, the more interesting the final effect.

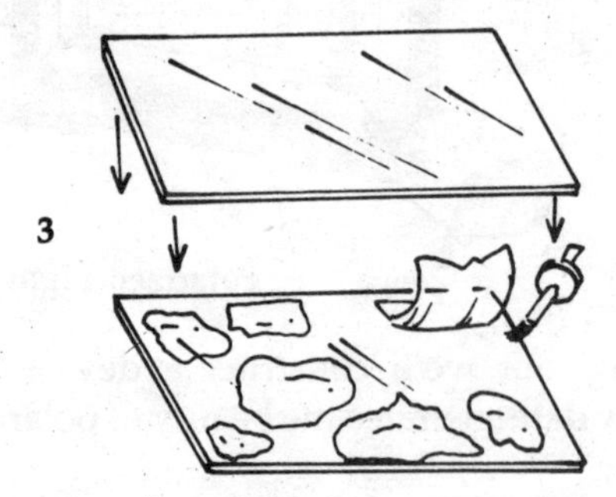

Place cover glass over cellophane glass.

Hold two pieces of glass together with tape (discard drawing).

Place a second piece of glass over your picture, and secure both pieces of glass together by stripping a little duct tape along the edges. Flip the taped glass so that the cellophane side faces down.

Attach the 12 × 16-inch (30 × 40-cm) plastic sheet polarizing filter to the glass facing you, securing it along the edges with duct tape. Make sure it sits flat against the glass. Cover the filter with a third piece of glass for a three-layer sandwich of clear glass, filter glass with cellophane picture, and clear glass.

Insert the sandwich into the back of the wooden picture frame. If the frame is deeper than the sandwich, secure the glass with balsa wood pressed under the lip and into the corners of the frame. If the sandwich fills the frame to the top, use carpenter's glue to attach the pieces of balsa wood over the corners of the frame. Allow the glue to dry.

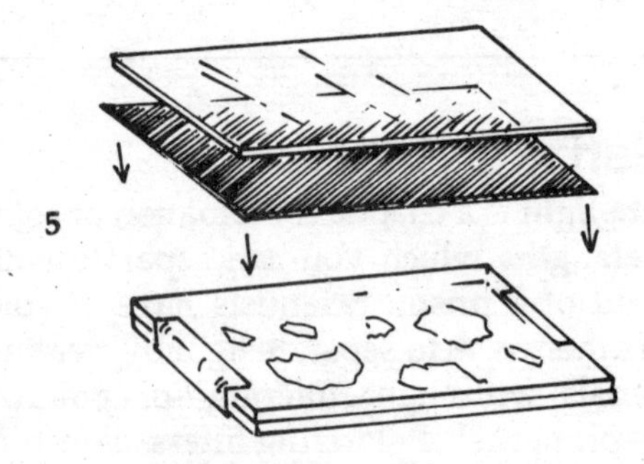

Add polarized filter and another sheet of glass.

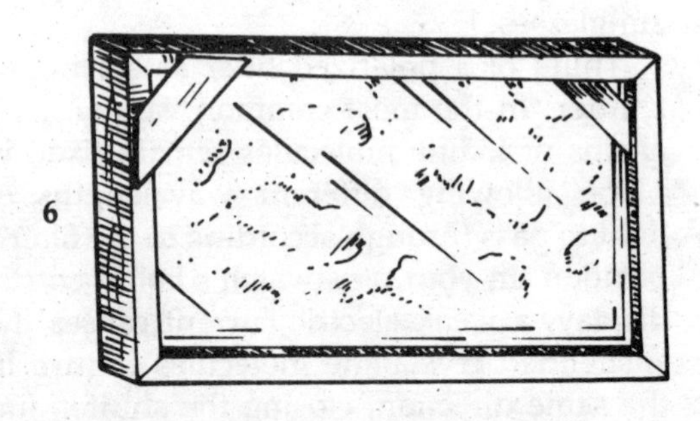

Mount glass in frame. Note two methods for securing balsa wood.

Viewing Disk

The next step involves constructing the viewing disk. With scissors, carefully trim the corners from the 5 × 5-inch (12.5 × 12.5-cm) sheet of polarizing filter until you have a reasonably accurate 5-inch (12.5-cm) diameter circle. Punch a hole in the center of the circle with a screw. Carefully screw the circle to the top of the wooden dowel, placing the washer between the filter and screw head. A vise can hold the dowel in place while you attach the screw. If a vise is unavailable, prop the dowel against a hard surface. Have a friend hold it with pliers to prevent it from twisting. Your friend should wear gloves as protection against any accidents.

Cut a 11⅞ × 15⅞-inch (29.7 × 39.7-cm) rectangle in the bottom. Place the box on its side, and carefully insert the picture frame into the rectangle, maneuvering it back and forth for a snug fit. Use duct tape around the front and back edges of the frame to secure it to the box. Paint the outside of the box black and the inside white.

Place the hooded lamp inside the box, and position the light so that it shines through the glass. From the outside, your cellophane picture is hardly impressive—merely a pale sheet of indistinct shapes. However, if you slowly revolve the viewing disk while looking through the polarizing filter, your picture swims with brilliant, shifting colors!

Collect polarized artwork to display in your box. You could have viewers not simply to view these polarized colors but make their own pictures, too.

Dancing Lights

You Will Need

- Round oatmeal box, 18-ounce (540 ml) size with removable plastic lid
- Two 4-inch (10-cm) diameter plastic funnels
- Aluminized Mylar from a helium balloon
- Rubber balloon
- 3 feet (90 cm) of old garden hose (or flexible tubing)
- Stiff cardboard, 9 × 12 inches (22.5 × 30 cm)
- Large sheet of white poster board or drawing paper
- Flashlight
- Rubber cement
- Duct tape

Visible Sound

Different sounds can produce different patterns of vibration on an elastic surface covered with reflecting material. These reflections, in turn, allow us to see the vibrations as patterns of light on a wall.

Cut off the oatmeal box's bottom. Cut the rubber balloon in half and stretch the top piece over an open end of the cylinder. To do this, tape one section of the balloon to the cylinder's edge and stretch the rubber to the opposite side, taping it securely. Then, move around the cylinder's edges, taping the balloon on one side, then stretching, and taping it on the opposite side until you create a tight drum-like surface. Secure the balloon by wrapping tape twice around the cylinder's edge.

Cut the Mylar into three circles with ¼-inch (.62-cm) diameters. With rubber cement, attach the circles to the drum surface in a triangular formation. Place each circle equidistant between center and edge.

Insert funnels at both ends of the hose. If the funnels are too small and fall out of the hose, wrap duct tape around the spouts to act as spacers. If the spouts are too big, make a small lateral slice in the hose to expand it, wrapping tape around the joint. Then tape a funnel to the cylinder's open end.

Dancing Light Setup

Stands

To construct stands for the cylinder and flashlight, cut out the patterns shown from cardboard. Discard the cut-out triangular pieces, and partially fold each larger piece down the center. Use the piece with the deeper cut to cradle the cylinder drum and the shallow-cut piece for holding the flashlight. To prevent the drum and cylinder from shifting, attach tape to the ends that touch the flat surface.

Place the flashlight and stand 1 foot (30 cm) in front of the drum, adjusting the light beam so that it falls evenly on the drum's surface. Behind the flashlight and directly opposite the drum, hang the white poster board. If you darken your immediate area a little, you'll see three reflected dots from the Mylar.

Position yourself behind the drum, and carefully lift the end of the hose with the funnel. Talk into the funnel and watch the reflections on the poster board. Notice how the patterns change as you lower and raise your voice and speak at different pitches. Then, position a small radio or cassette player by the funnel, and watch the sound reflections. You'll see dancing lights.

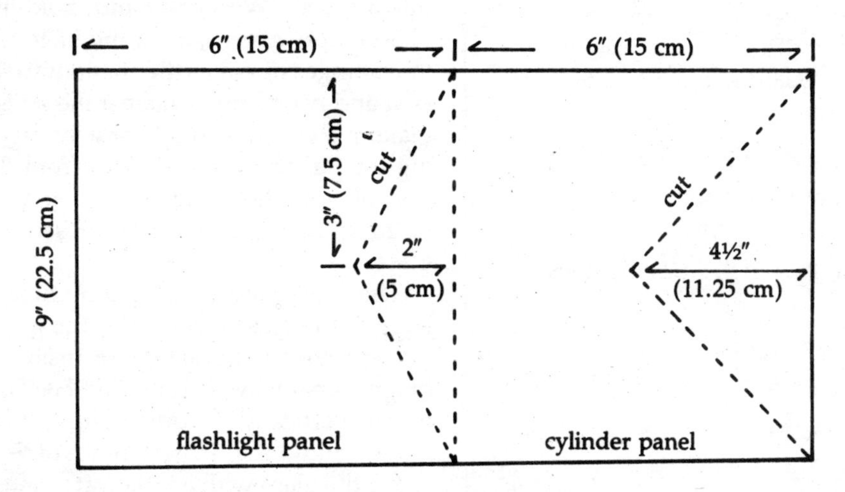

Pattern for Stand

Singing Glasses

You Will Need

- 8 wineglasses
- Wooden board about 1 × 3 feet (30 × 90 cm)
- Felt 1 × 3 feet (30 × 90 cm)
- 2 strips of thin wood 3 feet × 1 inch (90 cm × 2.5 cm)
- 2 strips of ¼-inch (.62-cm) thick foam
- Carpenter's glue
- Wood screws
- Wood stapler
- Vinegar
- Small saucer

Sound Waves

Sound waves—like waves of light and electric current—travel through various substances with different degrees of success. The important distinction between **waves** and **electrons** has to do with **frequency. Higher frequencies** of light waves create the blues and violets of the spectrum, and **lower frequencies** create the reds and oranges. With sound, different frequencies of wavelength create different **pitches**. And these pitches are the basis for the design of most musical instruments.

Pitch Test

First, try an experiment. Pour a little vinegar into a saucer. With one hand, hold the base of an empty glass against the table. Dip the index finger of your other hand into the vinegar and gently rub it against the glass's rim, making a circle. You'll hear a strange, ringing tone as the glass vibrates from the friction of your finger against it. If you hear nothing, apply a little less pressure to the glass.

The sound waves created by rubbing the edge of the glass resonate in the glass itself rather than in the air space inside. These longitudinal waves—so-called because they travel vertically through the glass—also move around the circumference of the glass.

Fill the glass with a little water and repeat the procedure. Add more water and try again. Notice how the pitch *falls* as you add water. Since pitch is determined by sound-wave frequency, we can say that adding liquid *lowers the frequency* of sound waves travelling through the glass. More water results in more drag on the glass—that is, more water molecules drag along with the glass molecules as they vibrate. This greater mass results in a lower sound speed.

Something Instrumental

Using this principle, we can construct a musical instrument out of glasses, with each glass tuned to a different note of the musical

Glass Harmonica

scale. Stretch felt around the wooden board and staple it to the back. Make sure you wrap corners neatly and avoid bumps where you staple. Trim the remaining felt to make the board sit level on the tabletop. To ensure this, you could buy small rubber furniture pads and screw them to the board's bottom corners.

Arrange the glasses neatly in a row, spacing them 1 inch (2.5 cm) apart. Cut the ¼-inch (.62-cm) thick foam into strips to fit neatly under the wooden strips. Glue the foam to the strips, and allow the glue to dry at least an hour.

Place the foam-backed strips lengthwise on the board, one strip on each side of the row of glasses. Move the strips towards the glasses until each strip just slightly overlaps the base of each glass. Carefully attach the strips to the board with wood screws at the ends, but avoid tightening the screws too much. You want the strips to act as cleats, securing the glasses against the board.

Tuning Your Instrument

To begin tuning your glasses, completely fill the glass to your left with water. Dip a finger into the vinegar and rub it. Keeping that pitch in mind, pour a little less water into the next glass in the row. You should hear a distinctly higher pitch from this second glass when you rub it. Continue tuning by adding less and less water to the glasses as you move down the row. You want to create an eight-note (octave) scale. Think of the first glass as "do," the second as "re," the third as "mi," and so on.

Now dip the index finger of each hand into the vinegar (keep two saucerfuls close by) and begin to play. Play tunes or harmonies with your singing glasses—a strange and beautiful music.

Bottle & Pipe Trombone

You Will Need

- Plastic watercooler bottle
- 2 feet (60 cm) of ¾-inch (1.8-cm) - diameter PVC (plastic plumbing) pipe

The Pitch

A bottle and pipe trombone demonstrates how the length of a vibrating air column affects pitch. Calling this a *trombone* might seem far-fetched, but like a real trombone, your bottle and pipe instrument produces different notes as you slide the tubing. And speaking of sliding, the sounds you make might remind you of another instrument you've probably tried, the slide whistle.

Trombone

Fill the watercooler bottle nearly to the top. Insert the plastic pipe as if it were a straw, with the straw's bottom resting at the bottom of the bottle. Hold the pipe upright and blow sideways against the top. With a little practice, you'll get a strong, high-pitched sound. The sound comes from a vibrating column of air inside the pipe. It's a small column at this stage, because most of the pipe is filled with water rather than air.

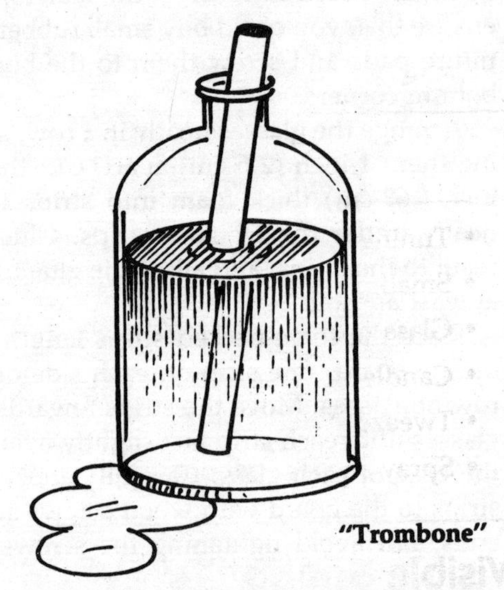

"Trombone"

Lift the pipe slightly out of the bottle, and continue blowing. What happens to the sound? Less water in the pipe means more air, and that means a *lower pitch*. Continue lifting the pipe from the bottle while blowing across the top. You hear a sliding series of pitches as the column of air changes.

With a little patience, you should be able to play the notes of a musical scale and memorize the position of the pipe for each note. A trombone player remembers positions the same way.

Sound Waves from a Tuning Fork

You Will Need

- Tuning fork pitched to A = 440
- Small piece of 30-gauge wire
- Glass microscope slides
- Candle
- Tweezers
- Spray shellac (optional)

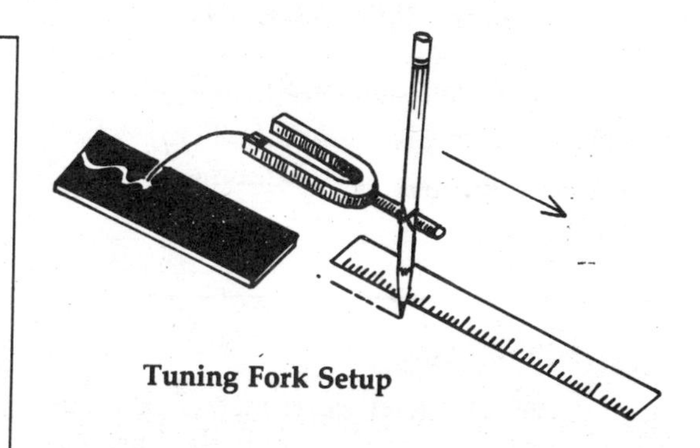

Tuning Fork Setup

Visible Sound Waves

Sound consists of **wavelengths** that we can sometimes see when an object vibrates.

For this demonstration, you'll need a candle. Be especially careful with an open flame, and thoroughly extinguish the candle before discarding it. With tweezers, grasp the corner of a microscope slide, and carefully pass it back and forth over the candle flame, just long enough to entirely coat one side with black soot. Prepare four or five slides this way.

Cut a 3-inch (7.25-cm) piece of wire, and, with as little tape as possible, attach one end to a tuning fork prong. The wire should protrude from the fork in a curve, with the wire's tip about 2 inches (5 cm) below the prong when you hold the fork horizontally.

Attach the shaft of the tuning fork 2 inches (5 cm) from a pencil's bottom with a little wire or tape. Secure the fork at a 90-degree angle, or as close as you can get. Position both ruler and glass slide horizontally. Secure the slide flat against the table surface with tape, looped in the back. Place the top right corner of the ruler against the bottom left corner of the slide.

Delicate Balance

Grasp the tuning fork by the shaft and strike it against the table's edge. Immediately slide your grip down to the pencil, holding it straight so that the fork prongs stick straight out, and the tip of wire touches the top of the slide. With your left hand holding the ruler firmly in place, drag the pencil along the ruler's edge so that the wire moves steadily over the slide, scratching a wavy line on the black surface. The wavy shape indicates that the tuning fork is vibrating at a pure pitch.

Collecting Sound Waves

Repeat the procedure, pulling the wire across the slides at different speeds and noting changes in the waveforms. Strike the tuning fork with less force, too, and notice the difference. The harder you strike the fork, the more it vibrates, and the greater the distance between the wave's peak and valley.

Make a collection of sound-wave slides, using various sizes of tuning forks. If you apply a thin coat of spray shellac to each slide, you can preserve your sound waves for display.

Echoing Hose Phone

You Will Need

- 100-foot (305-meter) old rubber garden hose
- 2 large funnels
- Duct tape

Echoing Hose Phone

Sound Waves

Sound waves travel very fast—744 miles (1,198 km) per hour in cool, dry air. Although it seems impossible to imagine something moving at that great a speed, you can under certain conditions hear sound waves "bounce" back to you after travelling great distances. The echo of your voice from a distant canyon is one example.

You can actually calculate distances by measuring the time it takes a sound to echo back to its source. Certain animals, like bats, use the reflection of sound waves to help them navigate. Finding your way around with sound is called **echolocation.**

Sound Experiment

To experience the speed of sound, begin by carefully cutting off the metal couplers at the ends of an old garden hose. To do this, you may need a saw or sharp knife, so ask an adult for help. If you cannot find a hose 100 feet long, use several hoses joined together.

At each end of the hose, insert the narrow end of a funnel. The funnels should fit tightly in the hose ends. Secure them with duct tape.

The rest is easy. Sit at a table with two funnels in front of you. Stretch the hose out away from you, looping and twisting it as much as possible. Hold one funnel to your mouth and the other against your ear. Now whistle or give a short yell and wait. In less than a second you'll hear your voice come back to you through the other funnel.

What happened? Let's do a few simple calculations. If sound travels 744 miles (1,198 km) per hour, divide that number by 60 to find that sound travels 12.4 miles (19.9 km) per minute. To find the distance sound travels per second, first convert 12.4 miles into feet (or 19.9 km into m). Since 1 mile equals 5,280 feet, 12.4 miles equals 65,472 feet. Dividing 65,472 feet by 60 (for seconds) yields 1,091.2—nearly 1,100 feet per second. Since the hose is 100 feet long, your voice comes back to you at a delay of about 1/10 second. That doesn't seem like much, but it's just enough for your brain to recognize.

Let's try that in metrics. If sound travels at 1,198 kilometers per hour and almost 20 kilometers per minute, to find the distance sound travels per second, first convert 20 km into meters. Since 1 kilometer equals 1,000 meters, 20 km equals 20,000 meters. Dividing 20,000 meters by 60 seconds equals 333.3 meters per second. Since the garden hose is 305 meters long, your voice returns at a delay of about 1/10 second.

Speed of Sound

Knowing the speed of sound can be very useful. The next time you're in a place where your voice echoes, time the echo with a stopwatch. By reversing these calculations, you can figure out the distance your voice travels before it returns to you.

Boom-Box Tube

You Will Need

- 4-foot (120-cm) plexiglass tube 5-inch (12.5-cm) diameter
- 6-inch (15-cm) square plexiglass piece
- Plexiglass glue
- 5 cups (1,200 ml) of plastic-foam beads
- Measuring cup (1 cup = 240 ml)
- 2 pieces of plywood for trestles 6 × 8 inches (15 × 20 cm)
- Boom box radio or tape player with microphone attachment

Good Vibrations

Sound travels through the air in waves. However, unlike waves of light, sound waves create *pressure* as they move, squashing air molecules against each other before they expand again and pass their energy to neighboring molecules. Although individual molecules vibrate back and forth, they do not actually move through the air. This **vibration,** passing from molecule to molecule, is what we *hear*.

But you can *see* various configurations of sound waves by placing a thin layer of sand on a drum, and singing loud enough to make the skin vibrate. Or you can drag a vibrating tuning fork across a blackened surface. But a dramatic way to demonstrate sound waves in action involves showing how they collide and reinforce each other in a closed space.

Collect enough plastic-foam beads to fill about 5 cups (1,200 ml). You can carefully break apart a piece of plastic-foam packing material. Or you could use an old beanbag chair—have an adult carefully open it along the seam, then sew it together again.

Prepare the plexiglass tube by gluing the square to one end. Allow the glue to dry thoroughly before handling it.

The trestle height depends on the position of your boom box. Place the open side of the tube against the speaker, centering it.

Turn the plywood pieces narrow side up, and make a shallow, wedge-shaped cut in the top side of each piece. To determine trestle height, measure from the center of the speaker to the tabletop. Cut the appropriate amount of wood from each piece and nail it to the trestle's bottom.

Turn the tube on its end, and add the plastic-foam beads. While stopping the

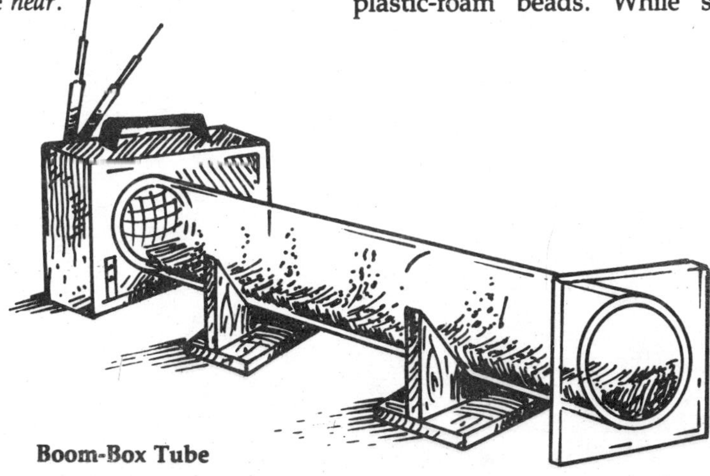

Boom-Box Tube

open end with one hand, turn the tube sideways, and shake the beads so that they spread evenly. Straddle the tube across the trestles, and push the boom box speaker against the open end.

Testing—1, 2, 3

Insert the microphone into the boom box, and sing, whistle, or make any other sustained sound at various volumes and pitches. Observe the beads' movements inside the tube. With every sound you make, particularly of sustained high and low frequencies, beads will jump and line up in balanced ridges or wave-like formations. Adjust the volume of the boom box to see these formations clearly.

The movement of the beads indicates behavior by sound waves in a closed space. As the boom-box speaker moves back and forth, it sends waves of compressed air racing down the tube. When these waves reach the end of the tube, they bounce off the closed side and travel back towards the speaker, where they meet with waves travelling in the opposite direction. If the waves are the same frequency in both directions, they reinforce each other and create a stable, high-pressure formation called a **standing wave.** You see the outline of this formation as beads move away from the peaks of the standing wave and settle into low-pressure troughs.

Replace your voice with music. Though you'll probably see fewer standing wave formations, the dancing beads inside the tube should fascinate your viewers.

Edison's Reproducer

You Will Need

- Portable phonograph
- Old record
- Frozen-juice can
- Empty matchbox
- Aluminum foil
- Small cork
- 3 sewing needles
- Nail
- 2 large pieces of poster board (one white)
- Rubber cement
- Ballpoint pen cartridge

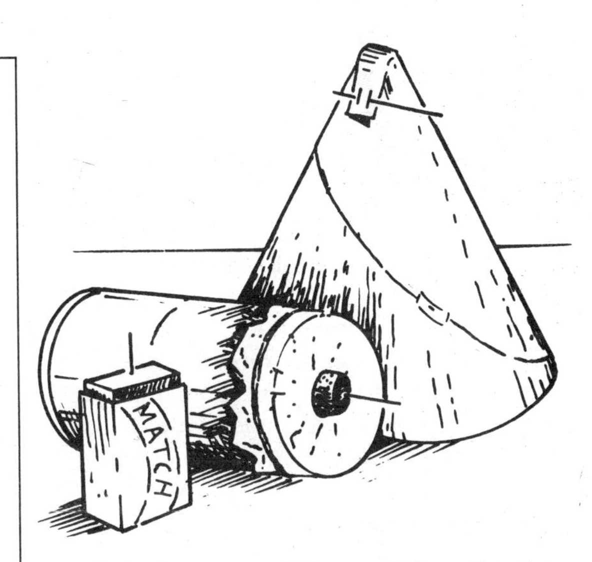

Sound Device

Just as sound from a tuning fork produces patterns on glass, patterns on a record surface produce sound. But to hear the sound from a record surface, we need something to make the sound.

Thomas Edison recognized this problem in early experiments with the talking machine. His solution was the phonograph pickup, or **reproducer,** a simple device that **resonates,** or makes small vibrations more noticeable to our ears.

Edison's reproducer eventually became a sophisticated instrument made from a thin sheet of mica in a carefully designed box. Soon, he added the familiar horn to thrust the sound outward. You can construct several working models of reproducers that demonstrate basic principles of sound. Place an old record on the turntable and lightly touch the surface with a needle. Do you feel vibrations in the needle? If you place your ear close to the record while holding the needle, you hear thin, scratchy music.

Slide the cover off the matchbox. Starting from the inside of the matchbox, carefully insert the needle halfway through one of the narrow edges. Slide the cover back on, and, while holding the matchbox by the sides, gently touch the needle against the record again. Notice how much louder the music becomes with the matchbox acting as the reproducer.

For a more effective reproducer, remove both ends of an empty frozen-juice can with a can opener, and press any sharp edges against the sides. Stretch aluminum foil across one opened end and secure it with a rubber band to make the foil tight and drum-like. Carefully tap the needle through the cork until the eye-end is flat with the cork's surface. Attach the cork to the center of the aluminum foil with rubber cement, and allow a half hour for the cement to dry.

Gently touch the needle to the record surface while holding the can. Note the better quality of sound. This model is actually a reproducer and horn design in one; the juice can acts as a horn.

Finally, roll the non-white piece of poster board into a cone by starting along one of the corners. Tape the cone together, and trim the wide side with scissors. Fold the narrow end of the cone and poke the needle through it, so that most of the needle protrudes at a 90-degree angle from the cone.

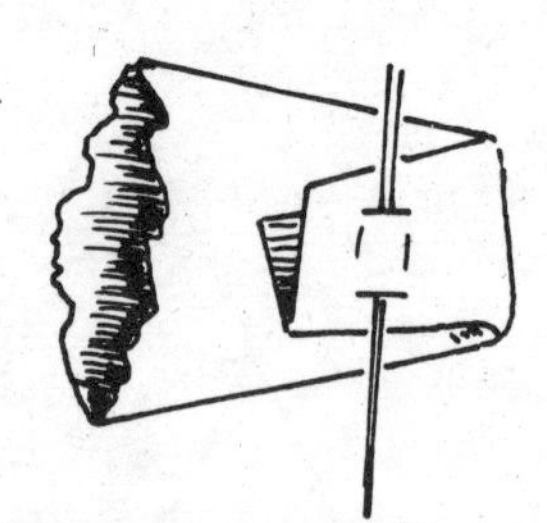

Detail of Needle through Cone

Hold the cone along the sides and gently guide the needle on the record. The sound you hear should be of reasonably high quality, and it probably reminds you of an old-fashioned wind-up phonograph recording!

Sound Reversal

You can demonstrate the principle of sound reproduction reversed or how Edison discovered that live sound could be captured on a soft, waxy surface and preserved forever.

Cut the white piece of poster board into a smaller square and punch a hole near the center. Place the poster board on the turntable so that it spins at a low speed. Hold the pencil against the poster board close to the turntable's edge (you can estimate this) and let the revolving turntable do the rest. Remove the poster board from the turntable and trim it along the pencilled circle to create a perfectly balanced paper "record."

Carefully remove the needle from the cork of the juice can reproducer. Place one hand inside the can, and press a finger against the foil, supporting the cork from the back. With the nail, carefully widen the hole in the cork so that you can insert the ballpoint pen cartridge.

See Your Own Voice

Place your record back on the turntable and hold the pen side of the reproducer against it. Notice the plain, spiralling line. Begin talking or singing into the open end of the reproducer, and watch the line explode into squiggles and peaks and waves that describe your voice!

Whispering Balloons

You Will Need

- 2 large latex (or similar) balloons
- 2 large plastic foam boards
- Craft knife
- Straight-edge or metal ruler
- Stiff cardboard

A Matter of Density

The design of musical instruments shows us how sound waves, like light waves, can travel through various shapes with different results. White light passing through a prism produces color. Sound passing through a variable column of air produces pitch. The slide whistle produces a higher pitch as you push the slide towards you, shrinking the air column inside the tube. A filled wine-glass vibrates at a lower pitch than an empty one because the sound waves are slowed down by the greater mass.

But do sound waves actually travel better through certain substances? Why do distant sounds seem closer on foggy nights? Why do your ears pick up the smallest sounds when you swim underwater? The answer is **density**. Water is *denser* than water vapor, and so it is the best conductor. But water vapor is a better conductor than dry air.

Gases also vary according to their densities, and some make better conductors of sound that others. Ordinary air is a mixture of gases and a good conductor of sound, but pure carbon dioxide is much denser and a superior conductor.

Balloon Setup

You can demonstrate the difference by filling two balloons—balloon A with air from a bicycle pump and balloon B with air from your lungs. After you fill the balloons, place each on a plastic foam board and carefully trace the shape with a felt-tip marker. Cut holes in the boards large enough to contain the balloons, making sure each fits snugly in its hole, using tape if necessary. If a balloon is loose, discard it and inflate another.

To keep the foam sheets with balloons standing upright, construct a simple brace. Cut slits near the edges of each sheet; then, cut the stiff cardboard into four strips. Insert two strips crossways into the slits of each sheet.

Balloon Test

Place the sheets side by side on the table. Sit at the table and have a friend sit opposite

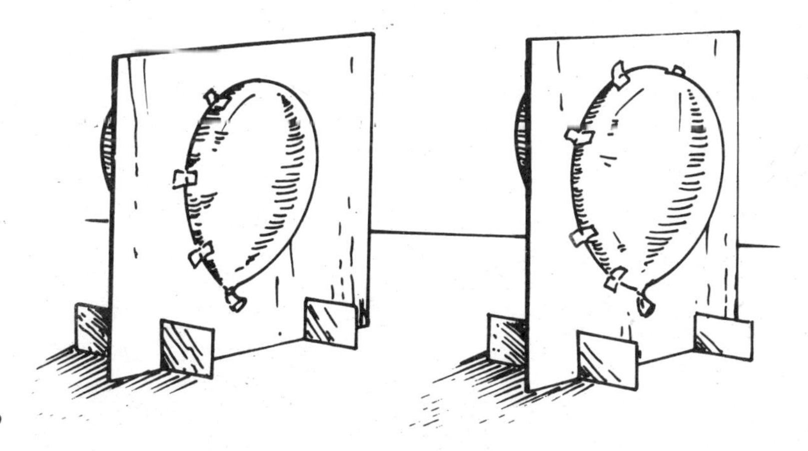

Balloon Setup

you so that he or she is completely blocked from your view by the sheets. Both you and your friend should have a piece of paper listing 10 items or phrases to whisper into balloons A and B alternately.

Person #1 indicates to person # 2 that he is about to begin by raising a hand over balloon A. He then places his mouth close to the surface of the balloon and whispers the first item or phrase on his list. Person #2, keeping an ear close to the balloon, writes it down. Continue back and forth—one person whispering while the other records—until the whisperer completes all the items on his list. Then let the listener do the whispering.

Results

Compare lists for accuracy. How many words or phrases heard through balloon A were accurate? How many through balloon B? Notice how words or phrases whispered through balloon B were much clearer than those whispered through balloon A. Notice, too, how sound coming through balloon B seemed more focused and concentrated, whereas sound coming through balloon A was diffuse and weak. Can you guess how air in balloon A differs from air in balloon B?

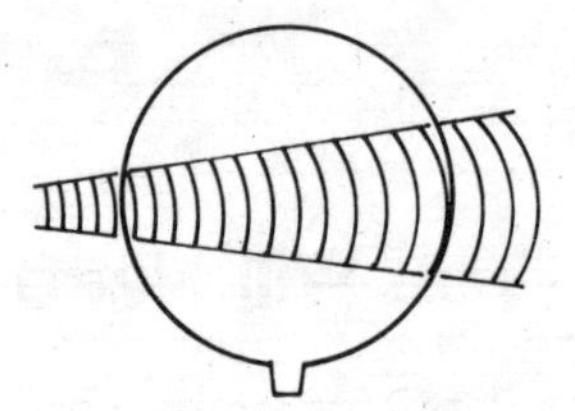

Sound Waves through Air

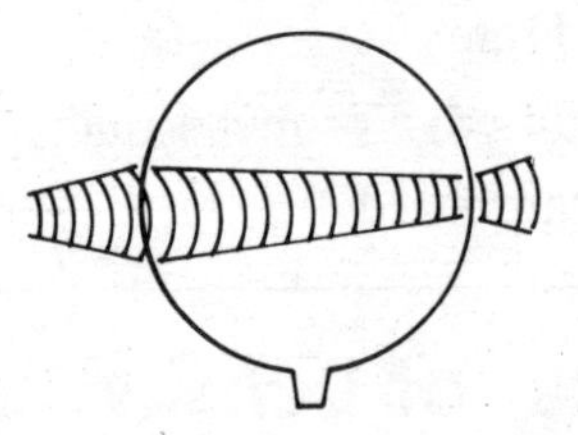

Sound Waves through Carbon Dioxide

The ordinary air from a bicycle pump is less dense that air from our lungs, which is mostly carbon dioxide. Sounds travel differently through balloon B, bending and concentrating on the opposite surface, rather than spreading out and weakening as it does in balloon A.

Parabolic Sound-Collecting Dish

You Will Need

For Dish and Frame

- Iron wok or similar bowl, about 15¾ inches (39.37 cm) in diameter
- 5-inch (12.5-cm) diameter circular fixture box for hanging lamp with 1-inch (2.5-cm) diameter center hole
- Lightweight microphone 5 × 1 inches (12.5 × 2.5 cm)
- Cassette or tape recorder (compatible with microphone) that can be monitored with earphones
- 1 × 2-inch (2.5 × 5-cm) wood—2 pieces 10 inches (25 cm) and 1 piece 17½ inches (43.75 cm)
- ½-inch (1.25-cm) thick neoprene rubber disk, 5-inch (12.5-cm) diameter
- Threaded steel rod 15½ × ¼ inches (38.75 × .625 cm)
- Wire frame from small lampshade, 3-pronged, 4 inches (10 cm) long
- 2 cabinet hinges 3 inches (7.5 cm) long
- ⅜ inch stove bolts and nuts for cabinet hinges
- 2 nuts ¼ inch square and 2 wing nuts ¼ inch
- 2 large fender washers 1½-inch diameter and 2 small washers ½-inch diameter
- Two 1-inch (2.5-cm) nails
- 4 corner braces, each with 4 screw holes
- 16 short wood screws
- 2 stove bolts for pivot seal 2 × 10/32 inches and 2 nuts to fit
- 2 rubber washers for pivot seal 1½-inch diameter × ¼ inch

For Platform

- 7-inch (17.5-cm) diameter plywood disk
- 1½ × ¼-inch (3.25 × .62-cm) threaded rod (sawed off longer piece)
- ¼-inch wing nut and washer for threaded rod
- ¼-inch coupler for threaded rod
- Photographer's tripod with standard ¼-inch (.62-cm) threaded rod for attaching camera

Additional Materials

- Power drill with ¼-inch and 10/32-inch bits
- Grinding wheel attachment for drill
- Hacksaw with fine-tooth metal cutting blade
- Level
- Fabric rivets
- Eyedropper
- Red food coloring
- Wide bowl or flat-edged candy dish
- Clear lacquer (if iron wok used)

Parabolic Sound-Collecting Dish

Collecting Sound Waves

Sound waves can be bent, focused, and manipulated in various ways. You can apply these principles in the **parabolic sound-collecting dish.** This instrument will amuse you for hours. You'll be able to zoom in on anything from treetop bird calls to distant conversations. Connecting the instrument to a portable tape recorder will also allow you to collect sounds. And the dish is useful for the study of echoes.

Constructing the Dish

You don't have to use a wok for the dish component of this project. Any shallow bowl—such as a saucer sled or photographer's light reflector—will work, as long as the material is rigid. Of course, you'll have to adjust the dimensions of the bracket to fit. The larger the dish, the greater its sound-gathering power. But whatever type of dish you choose, avoid using hemisphere-shaped bowls that are sometimes joined to make Christmas decorations. A hemisphere is not a parabola; it curves too deeply and will not reflect sound waves properly.

The wok presents both advantages and difficulties. The shallow bowl provides the perfect angle for reflecting and focusing sound waves, but woks usually come with riveted handles which must be removed before you can attach the dish to the frame. Also, the iron of the wok rusts easily, and you'll have to coat it with a clear lacquer.

To remove the riveted handles of a wok, attach the grinding wheel to the drill and grind the rivet heads until flush with the handle piece. Carefully pry the handles off with the screwdriver and the rivets will drop out too. Use fabric rivets to close the old rivet holes, snapping them into place.

Levelling the Wok

Turn the wok with its open side facing you, and place it on top of the bowl or candy dish. If your level is too short to straddle the wok, rest it on a narrow piece of wood that's long enough to bridge the span.

Manipulate the wok until level (the bubble on your level will center itself between the lines), then carefully lift the level and turn it 90 degrees before setting it down again. Adjust the wok until level again. You can spot check the overall level by lifting the level bar and turning it 45 degrees, but be extra careful that you don't bump the wok as you move the bar around.

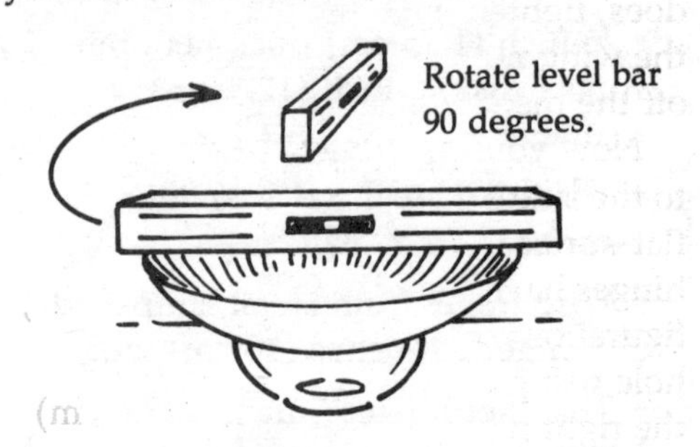

Levelling the Wok

When confident your wok sits level, remove the level bar, and mix red food coloring in a small amount of water. Using the eyedropper, carefully dribble some colored water down the side of the wok until you have a small, but clearly visible puddle at the bottom, no more than ⅛ inch (.31 cm) in diameter. Give the water a minute or two to settle, or tap the sides of the wok gently. The position of the puddle indicates the wok's exact center and where to drill a hole for the long threaded rod.

Carefully dip a #3 pencil through the puddle so that you can mark the center with an X or large dot. Remove the wok from the bowl, spill out the water, and dry it thoroughly before proceeding. If you choose to lacquer the wok, this is the time to do it. In a well-ventilated place, lacquer both sides with at least two coats and allow them to dry. In the meantime, use a hacksaw to cut 1½ inches (3.25 cm) from the threaded rod, placing the small piece aside for later.

Insert the ¼-inch bit into the drill, and, while having a friend hold the wok firmly against a piece of wood, drill a hole through the pencil mark you made. Place a ¼-inch washer and wing nut on one end of the threaded rod, and push the rod through the hole in the wok. Then place the washer and wing nut on the other end of the rod. Make sure the rod moves freely through the hole. If it's too tight, enlarge the hole a little by re-drilling.

With the rod about halfway through the wok, tighten the wing nuts on both sides to anchor the rod in place. Hold the sound dish in front of you, and check to see if the rod pokes straight out and does not sag. If it does, tighten the wing nuts. Remove one of the wing nuts, and slide the rod completely off the dish.

Now you must attach the cabinet hinges to the side of the dish. Place the hinges on a flat surface, spine side facing you. If your hinges have screw holes in a triangular configuration on both plates, drill an additional hole using the 10⁄32-inch bit in the center of the right plate. This ensures a smooth pivot where the dish joins the frame.

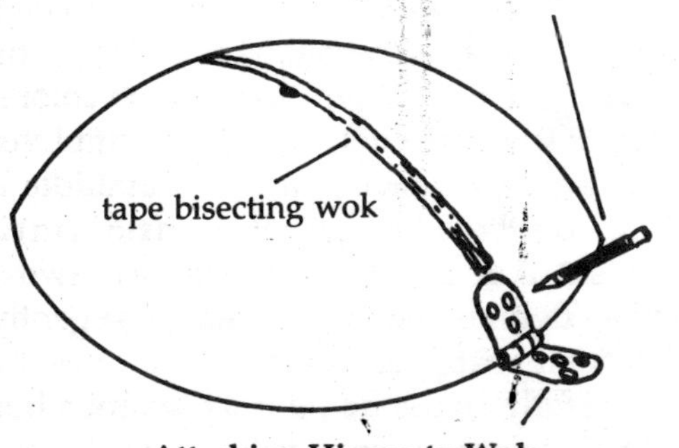

Attaching Hinges to Wok

Turn the dish over. You want to determine, and mark with pencil, two points along the edge directly opposite each other. Take a cabinet hinge, spine side facing you, and position the *undrilled* plate over each mark. Poke a pencil through the screw holes so that you know where to drill.

Using the 10⁄32-inch bit, drill over the screw hole marks. Attach the hinges to the back of the dish, poking the ⅜-inch stove bolts through the inside surface and securing them with nuts on the opposite side.

Use the pliers to tighten the nuts, but avoid tightening too much or you might bend the dish where the hinges attach. Remember, you want as smooth a surface as possible.

Constructing the Microphone Mount

Use the craft knife to cut a 1-inch (2.5-cm) hole in the center of the neoprene rubber disk. Although a craft knife cuts through rubber well, be very careful to hold the disk firmly when using this sharp tool. Push the microphone three-quarters of the way through, making sure you have a tight fit. Then remove the microphone.

Place the circular fixture box on a flat surface, hole side down. Squeeze both hands around the prongs of the lampshade frame, compressing them enough to fit into the open side of the canopy. Release the prongs so that they press firmly against the canopy's sides.

Turn the canopy on its side and press the neoprene disk over the prongs of the lampshade. You may need to make small slits in the rubber to allow the prongs to poke out. Finally, reinsert the microphone, wire first, through the neoprene disk and push it three-quarters of the way through. The back of the microphone should emerge through the 1-inch (2.5-cm) hole in the canopy. All that remains is to attach the mount to the threaded rod of the dish.

Remember, you left one wing nut and washer on the threaded rod. Turn the rod so that the end with the wing nut faces up (the washer might slide down the rod when you do this, but you can fix that later). While holding the rod in this position, twist a ¼-inch nut along the rod to about 1½ inches (3.25 cm) from the end. Place the fender washer over the nut.

Hold the microphone mount so that the narrow side of the lampshade frame sits against the fender washer. Then, cover the exposed side of the frame with another fender washer, securing it tightly with a nut. The mount is now "pinched" into place between two fender washers. Having attached the mount to the threaded rod, put the entire component aside.

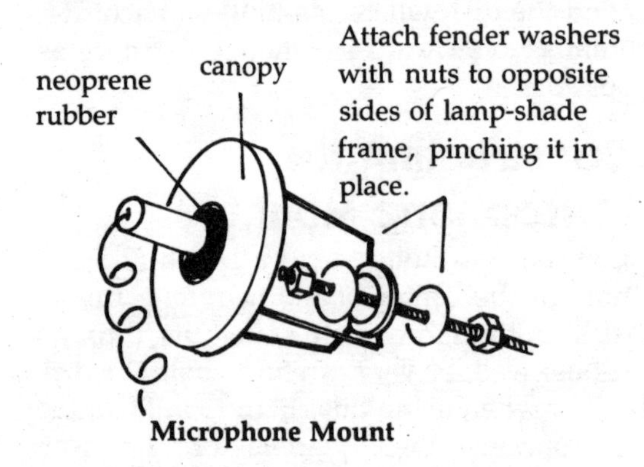

Microphone Mount

Constructing the Frame & Platform

The frame is relatively simple to construct. Measure 1 inch (2.5-cm) from the end of each 10-inch (25-cm) piece of 1 × 2 wood, and drill a hole with the 10/32-inch bit. Make sure the hole is centered on the width. Next, determine the midpoint of the 17½-inch (43.75-cm) piece of wood and drill a hole with the ¼-inch bit, centering it as before.

Turn the shorter pieces hole side down, and place the longer piece over them like a trestle. Have someone steady the wood as you nail the longer piece into the shorter pieces. To secure the frame, attach the four corner braces with wood screws where the pieces of wood join.

For the platform, buy a circular, 7-inch (17.5-cm) diameter piece of plywood from a lumberyard, or use the pencil compass to draw a circle on plywood scrap and have it cut out. At the center, drill a ¼-inch (.62-cm) hole. Hammer the ¼-inch coupler into the hole, taking care that you don't damage the coupler as you pound it through. Since most couplers have a hexagon shape, you may need to widen the ¼-inch (.62-cm) hole by repeated drilling. The platform should contain the coupler neatly, so that one end of the coupler is flush with one side of the platform.

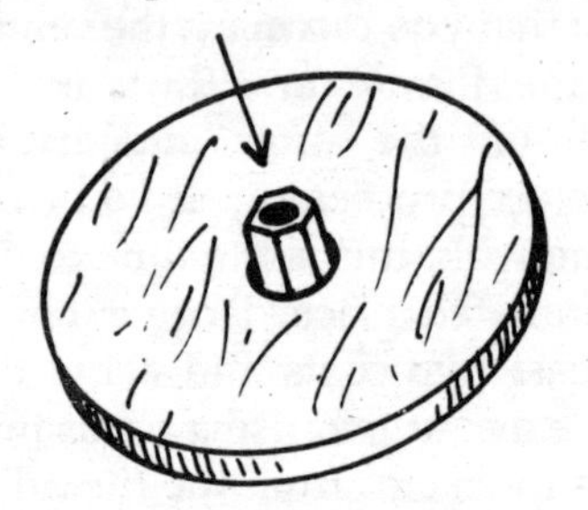

Platform Detail

Putting It All Together

You're now ready to assemble your sound dish and put it to work. Open the tripod, making sure to lock all joints. Your tripod should have the standard ¼-inch threaded rod for attaching a camera. Turn the platform flush side down, and screw the tripod rod into the coupler until the platform sits securely on the tripod. Now, take the 1½-inch piece of threaded rod (you removed it from the longer piece), and screw it into the coupler at the top of the platform. Place a rubber washer over the threaded rod.

Turn the dish hollow side down, and position the frame around it so that the ends of the frame align with the hinges. Starting at one side of the dish, slide a 2-inch (5-cm) stove bolt through the center hole on one of the hinge plates (the one you drilled), and insert a washer. Then poke the stove bolt through the frame hole, capping it off with a metal washer and wing nut. Repeat this procedure on the other side of the dish. When both sides of the dish are attached to the frame, tighten the wing nuts to keep the dish from swinging freely.

Carefully lift the frame and dish and place it on the platform so that the threaded rod pokes through the center hole of the frame. Where the rod emerges, drop a metal washer and a screw on a wing nut.

Attach the microphone mount to the dish by sliding the threaded rod—with attached microphone mount—through the hole in the center of the dish. Place a washer and wing nut on the rod, tighten it against the dish the way you did when you first tested the rod for straightness.

Some Ideas about Using the Dish

The dish is an excellent project for science fairs because you can easily demonstrate its operation to viewers. The wide space of a gymnasium or display hall provides many opportunities for selecting sounds. You could also have a tape handy of sounds you've already collected, like bird and animal calls. Echoes are particularly useful for testing the range of your dish. A previous project describes how you can calculate distance by timing your echo. You might categorize your echo collection based on distance alone, noting for your viewers the range at which a dish of this size exceeds its capacity.

To operate the dish, first plug the microphone and earphones into the tape recorder. Depending on the model of your recorder, you may not have to run the tape while monitoring the sound. Many recorders have a "PAUSE" feature which allow you to test sound levels before engaging the tape.

While wearing the earphones, point the dish towards a sound you'd like to pick up. Or maybe you don't have a particular sound in mind, but would like to pick out the conversation of a distant group of people. If the sound does not come in clearly at first, adjust the threaded rod by loosening the wing bolt at the back of the dish, then sliding the rod either towards or away from you. Your microphone must be adjusted to find the exact focal point of the sound. When you find that point, the sound will be crystal clear.

How It Works

The unique shape of a parabola concentrates and focuses certain sounds while filtering out others. When you point your dish towards a distant sound, those sound waves enter the dish in straight lines, bounce off the inside curve, and focus at a point near the microphone. Extraneous sounds that enter the dish from the sides, bottom, or top are bounced out of the dish and never reach the microphone.

You can see why the **parabola shape** is so useful for wave-collecting—from sound dishes to satellite antennae to giant radio telescopes, miles in diameter!

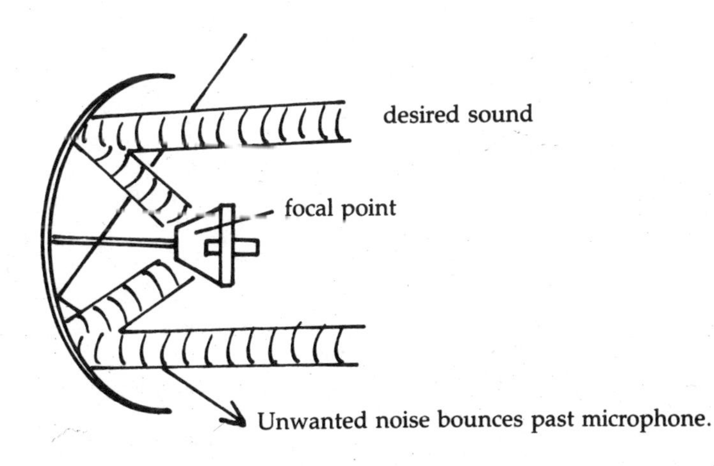

How a Sound Dish Works

How Does It Work?

Clay Boat
Boat-of-Holes
Diving Submarine
Measuring Friction
Air Cars
Why Airplanes Fly
Aerodynamics in a Wind Tunnel
Do Fans Really Cool You?
Hot-Air Balloon
Sturdiest Soap Bubble
Freestanding Arch
Stress Test for Bridges

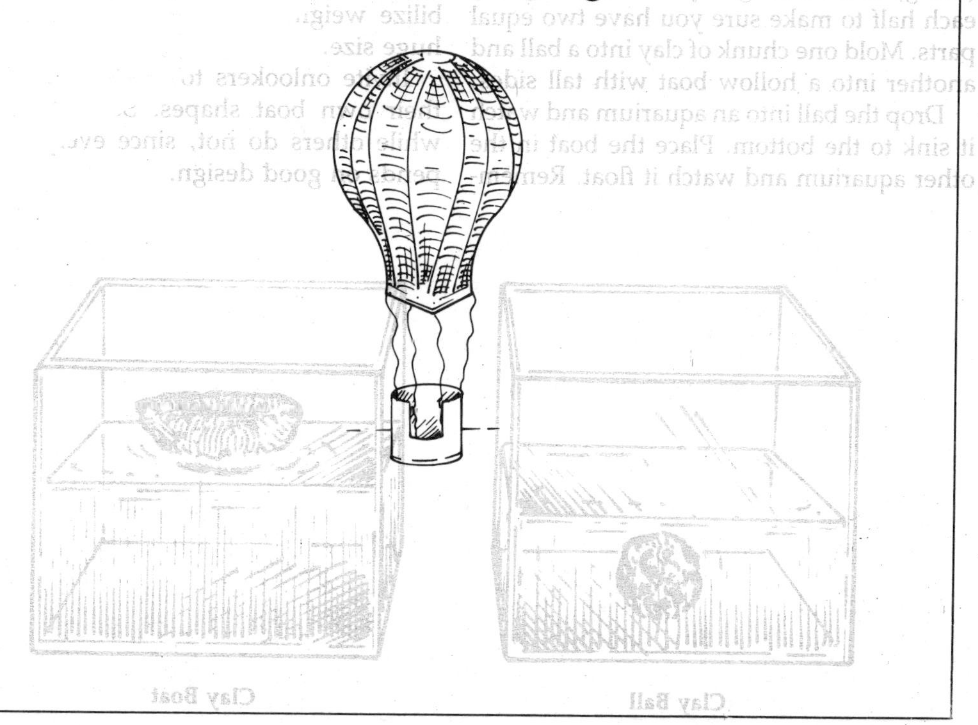

Clay Boat

You Will Need

- 2 aquariums, same size
- Blue food coloring
- 1 pound (500 g) of modelling clay
- Small scale

Sink or Swim

Ever wonder how a 200-ton (181-metric-ton) ocean liner manages to float? By using modelling clay and two aquariums, you can clearly demonstrate the principle of **buoyancy.**

Fill both aquariums three-quarters full of water. Mix just enough food coloring with the water to give it a bluish tint without clouding or darkening it. Divide the pound (500 g) of modelling clay in half, weighing each half to make sure you have two equal parts. Mold one chunk of clay into a ball and another into a hollow boat with tall sides.

Drop the ball into an aquarium and watch it sink to the bottom. Place the boat in the other aquarium and watch it float. Remember, both boat and ball were made from the same amount of clay. Do you notice anything different about the water level in the aquarium that contains the boat?

Remove the boat and squash it into a ball, dropping it into the aquarium again. It sinks! Remove the ball from the other aquarium and shape it into a boat. It floats! The water level is always higher in the aquarium that contains the boat.

Density & Displacement

Substances *denser* than water, such as metal or modelling clay in our experiment, sink because they weigh more than the upward thrust of water they *displace*. But dense substances can be made to float by reshaping them to increase the volume of water they displace, thereby increasing the upward thrust of the water. The hull of a 200-ton (181-metric-ton) ocean liner is carefully designed to maximize upward thrust and stabilize weight, so that it floats despite its huge size.

Invite onlookers to model the clay into their own boat shapes. Some boats float while others do not, since everything depends on good design.

Clay Ball

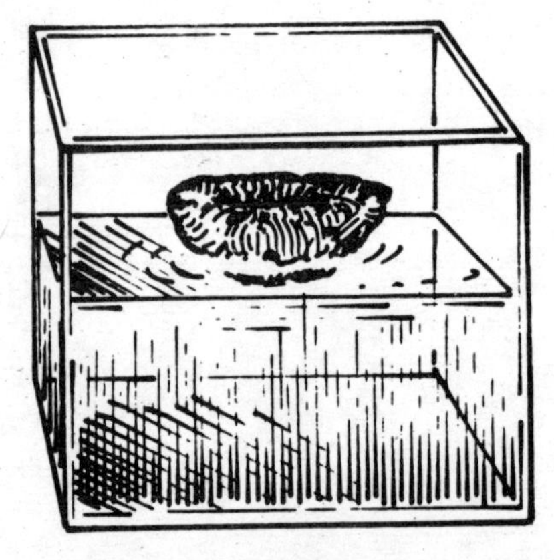

Clay Boat

Boat-of-Holes

You Will Need

- Piece of fine wire screen, about 4 × 4 inches (10 × 10 cm)
- ⅛-inch thick (.31 cm) sheet balsa wood, about 3 × 10 inches (7.5 × 25 cm)
- Shirt cardboard
- Craft knife
- Poster board
- Wooden matchsticks
- Matchbox
- Ball of absorbent cotton
- Denatured alcohol or rubbing alcohol
- Bar soap
- Aquarium tank or plastic wading pool
- Vegetable oil
- Eyedropper
- Epoxy glue

Surface Tension

Imagine water's surface as film made from tightly packed molecules. Certain objects made of nonbuoyant materials can float if you carefully place them in water. Smaller and lighter objects—like a dandelion seed—easily coast along the water's surface without even getting wet. You can demonstrate basic qualities of **surface tension** with model boats.

Boat-of-Holes

Using pliers, fold each corner of the fine screen towards the center, then fold each edge towards the center. You'll have a flat-bottomed boat with 1-inch (2.5-cm) sides and pointed corners. Carefully place this boat-of-holes flat side down on the water. Watch it float—held by surface tension alone.

Now gently tip one of the corners into the water, breaking the surface, and watch the boat sink like a rock. Remove your boat, dry it off, and try again. Why do you suppose the boat no longer floats? Replace the water for the next boat model.

Other Boats

With a craft knife, cut a 3 × 2-inch (7.5 × 5-cm) section from the end of the plywood. Trim this smaller section into an arrow-shaped boat that's flat at the stern (rear) and sharp at the bow (front). Cut a small notch into the stern, no deeper than ½ inch (1.25 cm).

Alcohol-Powered Boat

With epoxy glue, attach superstructures to the boat. Use a matchstick for a mast and poster board for a funnel. Allow the glue to dry thoroughly before testing your model.

Balsa-Wood Boat with Soap Engine

Cut a small chunk of bar soap, just large enough to fit comfortably into the wedge at the boat's stern. Carefully place the boat in the water and watch it move steadily forward, seemingly propelled by the soap engine. What happens is that the soap breaks the *surface tension* at the stern; so, the *elastic tension* at the bow pulls the boat forward.

A more elaborate boat model uses the larger piece of plywood, more poster board, a matchbox, cotton, and alcohol. Cut the balsa wood into the shape of a boat, no wider than the matchbox. Glue a ½-inch (1.25-cm) wide strip of poster board around the outside for the boat's sides. Remove the matchbox's inner part and glue it, open side down, to the center of the boat deck. Using extra glue, make sure the chamber behind the matchbox—or alcohol "fuel tank"— is sealed from the section in front of the matchbox. Attach a poster board funnel and matchstick mast for the boat model's detailing. Allow the glue to dry at least overnight.

When you're ready to test your boat, fill the rear chamber with a little alcohol. Shape a piece of cotton so that a few strands hang out over the back of the stern into the water. Carefully place the cotton in the fuel tank, and set your model down in the water. It should move forward at a reasonable clip, propelled by the alcohol drawn up into the cotton by **capillary action,** and out into the water where it breaks the surface tension.

For your last boat, cut the shirt cardboard into an arrow shape similar to the smaller balsa-wood boat. From the stern towards the center, make the cut (see illustration), using a craft knife or scissors.

Carefully float the cardboard in the water, and place a drop of oil over the small center hole. Again, your small boat appears propelled by its oil-drop engine. What actually happens is that surface tension at the bow pulls the boat forward. Oil, lighter than

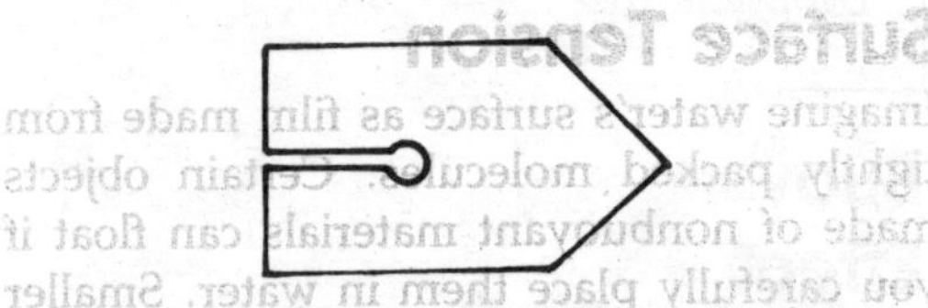

Cardboard Boat with Oil-Drop Engine

water, floats on the surface. As the oil runs out the opening at the stern, tension at the stern is reduced while the tension at the bow remains the same.

Demonstrate these miraculous boats, one at a time.

Diving Submarine

You Will Need

- Large aquarium tank or small plastic wading pool, filled with water
- ½-inch (1.25-cm) plywood 2 × 8 inches (5 × 20 cm)
- Long strip of tin plate 1½ × 22 inches (3.75 × 55 cm)
- 3 small squares of tin plate, one 1¼ × 1¼ inches (3.12 × 3.12 cm), two 1 × 1¼ inches (2.5 × 3.12 cm)
- 28-gauge (or narrower) uninsulated copper wire 1 foot (30 cm)
- 26-gauge (or thicker) uninsulated copper wire 4 inches (10 cm)
- 1-inch (2.5-cm) long copper tubing, just wide enough to fit over thicker wire
- Small, screw-in picture-hanging hook
- Hand drill with 3/32-inch bit
- 2 bolts with nuts 3/32-inch wide
- Metal clippers
- 15 brads
- 2 rubber bands 1/8-inch (.31-cm) thick
- Compass

Buoyancy & Bernoulli Effect

Submarines depend on both the **buoyancy** principle and the **Bernoulli effect** to accomplish what seems impossible—controlled sinking.

Submarine Construction

Draw a line laterally down the center of the long piece of plywood to guide your drawing the deck's shape. Make sure the deck is symmetrical by measuring along the vessel's length distances between the curved lines and the center line. Cut out the deck, keeping the remaining plywood scraps.

Mold the long strip of tin plate around the length of the deck, using pliers. You should have at least ½ inch (1.25 cm) of overlapping tin at the submarine's bow. Cut off anything extra. Attach the tin to the deck by hammering brads along the edge, spaced evenly.

On the 1¼ × 1¼-inch (3.12 × 3.12-cm) piece of tin plate trace, then cut out the design for the propeller, using the compass

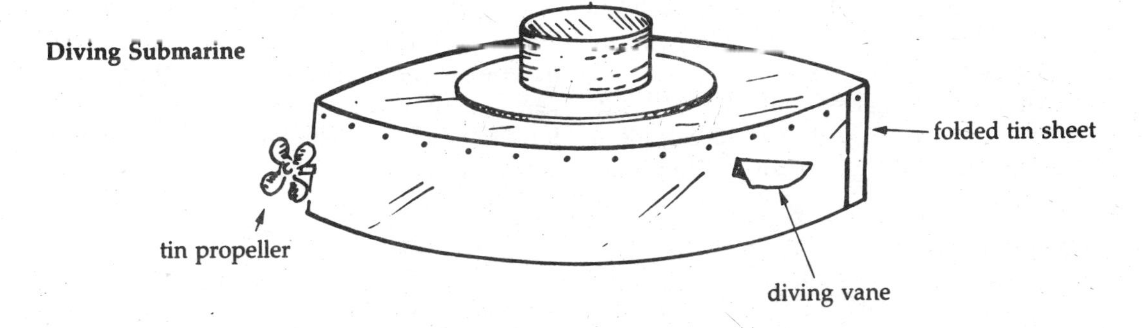

Diving Submarine

and metal clippers. In the propeller's center, punch a hole just large enough for the 26-gauge wire. Bend the end of the wire to keep the propeller from slipping off. To keep the blades free of the stern, slide the 1-inch (2.5-cm) piece of copper tubing over the shaft and against the propeller. Finally, use pliers to twist each propeller blade 45 degrees.

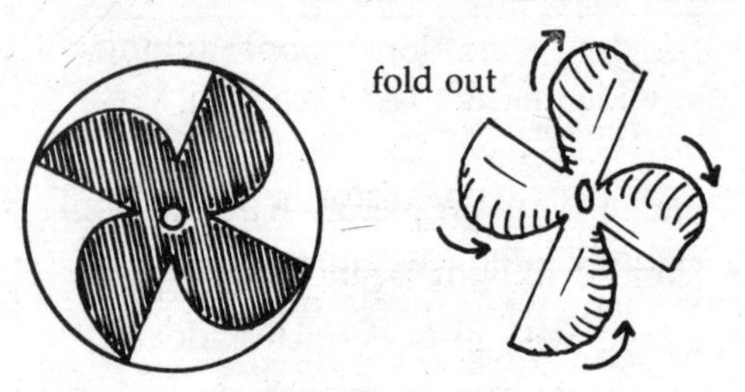

Propeller Pattern

Define the stern (rear) of the submarine by punching a small hole in the tin plate at one end. Turn the submarine plywood side down, and pass the propeller shaft through this hole, bending the other end into a hook for the rubber band. Center, then screw, the picture-hanging hook into the underside of the deck, 5 inches (12.5 cm) from the stern. Stretch the rubber band from shaft hook to picture-hanging hook, doubling it several times to make it taut. The propeller should sit snugly against the outside of the stern.

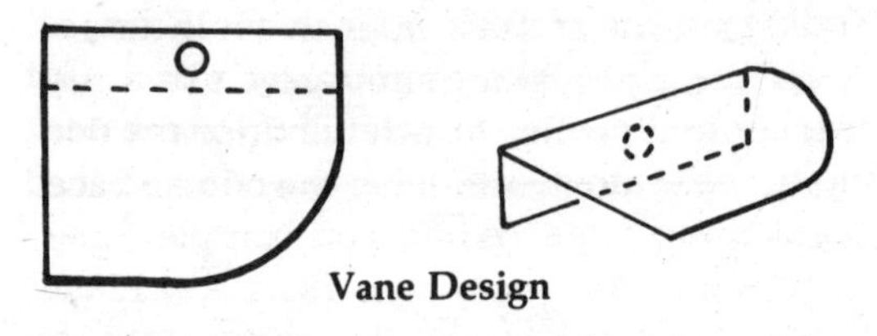

Vane Design

Two **diving vanes** act as wings for the submarine, allowing it to both dive and surface. Cut, drill, and bend each 1¼ × 1¼-inch (3.12 × 3.12-cm) piece of tin plate for the vanes (see pattern).

With the submarine still plywood side down, measure 2½ inches (6.25 cm) along each side from the bow (front). Then measure ¼ inch (.62 cm) from the plywood, and punch holes wide enough for the small bolts to pass through. Attach the vanes to opposite sides of the submarine, with bent sides facing you and curved sides facing the bow. Make sure each vane pivots freely.

Construct weights for the deck's underside by wrapping the 28-gauge copper wire around pairs of nails. Eight nails wrapped this way will provide four weights. Place one weight directly behind the bow, another directly before the stern, and the remaining weights on opposite sides of the deck. Use thumbtacks to secure them in place.

Turn the submarine right side up, and position the vanes with a slightly downward tilt. Finish your model by adding superstructures to the deck with plywood or tin scraps. But avoid overloading the submarine, or it won't glide through water smoothly.

Submarine Test

Gently place your submarine in the water. If weighted correctly, the deck should be at water level and slightly awash. If it sits below water level or too high, add or subtract weights as needed.

Remove the submarine and wind the propeller, holding it in place. Make sure the diving vanes remain at a slightly downward tilt, place the submarine in water, and let go. The submarine will move along the surface for a short time before going into a gentle, steady dive. When the propeller power runs out, the submarine floats to the surface.

Experiment with the angles of the diving vanes. Keep your submarine afloat by positioning the vanes parallel to the deck. Or create a steeper dive by tilting the vanes downward at a greater angle. Tilting the vanes too much, however, will increase water resistance, and your submarine won't be able to dive.

Measuring Friction

You Will Need

- Two 6 × 36-inch (15 × 90-cm) wood planks ½ inch (1.25 cm) thick
- Two 36-inch (90-cm) balsa-wood strips ⅛ × ⅛ inch (.31 × .31 cm)
- Wood block (small enough to fit inside plank)
- Thumbtacks
- 1 hinge with 4 screws
- Carpenter's glue
- Epoxy glue
- Sandpaper
- Clear plastic protractor
- Testing materials—aluminum foil, cellophane wrap, paper towel, sandpaper (fine, medium, and coarse), tissue, terry cloth, newspaper, shiny magazine cover, and plastic bag

Reducing Friction

Friction between solids can be reduced by keeping them apart with lubricants. But how much natural, or *dry*, friction exists between everyday surfaces, and how can we accurately measure it? To answer these questions, we'll construct an adjustable ramp and protractor measuring device.

Ramp & Measure

Sand one side of the first plank until the surface seems smooth and slippery. Place it, sanded side down, end to end with the second plank. Join the two ends with the hinge so that the sanded plank closes over the second plank like a book cover. If the hinge sticks, use machine oil on the joint.

Apply a very small thread of carpenter's glue to one side of each balsa-wood strip, and gently press the strips on top of the first plank, against the long sides. The strips should form a fence, keeping material from sliding off the plank's sides when inclined. After the glue dries thoroughly, use a red marker and a ruler to carefully draw a line along the seam between balsa wood and the plank's top edge. When you complete the

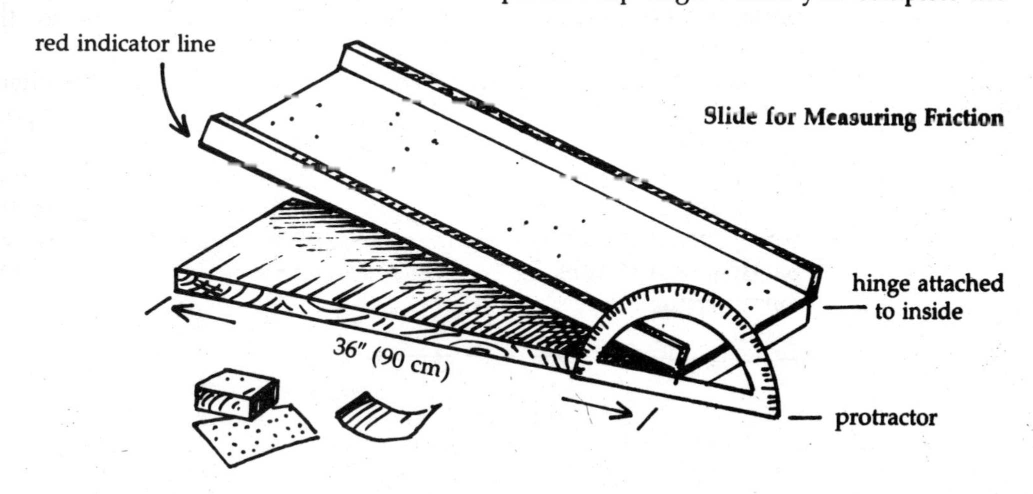

Slide for Measuring Friction

ramp, the line will show through the protractor and give you an accurate angle measurement.

With a small thread of epoxy glue, attach the plastic protractor to the side of the bottom plank so that the center of the protractor falls directly where the two planks join. Lift the ramp to make sure it clears the protractor and adjust it as necessary. If everything works, allow the glue to thoroughly dry.

Friction Test

The test involves placing the wooden block, wrapped in various materials, at the ramp's end and inclining the ramp gradually until the block slips down. At the point of slippage, an observer records the ramp's angle of inclination as indicated on the protractor. Repeat the test five times for each material.

The unwrapped wood block acts as a control. Begin by sanding it to the same smoothness as the first plank. Then, place it with its edge against the unhinged end of the ramp, and gradually lift that side of the ramp until the block slips to the bottom. If it catches on the way down, this does not invalidate the data. Record the angle, and repeat the procedure four more times.

Now, wrap the block with aluminum foil, creasing the foil around the top of the block only. Position the wrapped block against the ramp's edge, lift the ramp until the block slips, and record the angle of inclination. Again, repeat the procedure four more times, recording your data.

Continue by wrapping the block with plastic wrap, paper towel, sandpaper, tissue, terry cloth, newspaper, a magazine cover, and a plastic bag, successively. In each case, fold the material around the block and fasten it to the top with thumbtacks. Remember to record all data.

Angle of Results

The results, averaged on the chart below, may surprise you.

Why do you suppose less friction exists between aluminum foil and wood than between tissue and wood? What do the more slippery materials have in common? Do you think examining these materials through a microscope might help you arrive at some conclusions?

Friction Test

Material	Ramp Angle in Degrees
Aluminum foil	10
Shiny magazine cover	18
Plastic bag	20
Wood block	22
Plastic wrap	24
Newspaper	28
Sandpaper (fine)	36
Paper towel	39
Sandpaper (medium)	42
Sandpaper (coarse)	44
Terry cloth	46
Tissue	46

Air Cars

You Will Need

- Large balloon
- 2 L plastic soda pop bottle
- Craft knife
- Sandpaper
- Old record
- Plastic spool
- Carpenter's glue

Reducing Friction

Lubricants reduce friction between solids. Adding oil to a car engine, for instance, keeps the parts from wearing down by rubbing against each other. A simple air car demonstrates how a cushion of air also acts as a lubricant and reduces friction.

The Cutting Edge

With the ruler, measure 5 inches (12.5 cm) up from the bottom of the plastic bottle. Mark the spot with a grease pencil. On the opposite side of the bottle, repeat the procedure, so that you have two marks directly opposite each other, 5 inches (12.5 cm) from the bottle's bottom. Tie the string around the bottle so that it's level with the two marks. With horizontal strips of tape, attach the string securely to the bottle's surface.

Using the taped string as a guide, carefully cut around the bottle with a craft knife, making sure the cut edge is even and your car will sit level on the surface. Separate the bottle spout (which you'll use) from the rest of the bottle (which you can discard). Smooth the cut edge of the spout bottle with sandpaper, if necessary. Remove any remaining pieces of tape and string.

Inflate the balloon, and pinch the neck.

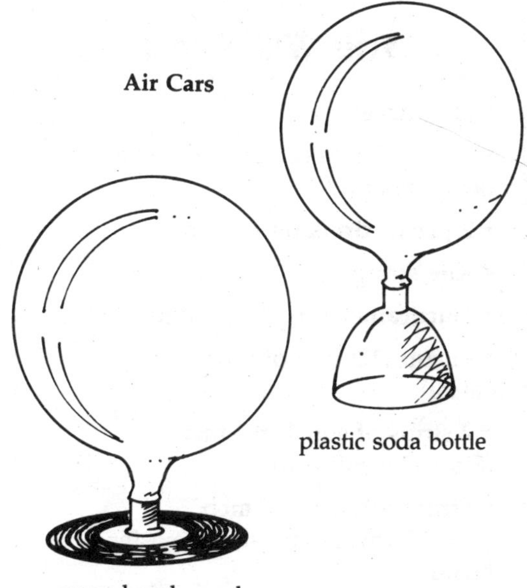

Slide a paper clip there to act as a clamp. Stretch the lower part of the balloon neck over the spout of the bottle, keeping the clamp in place, and put the air car on a smooth surface.

Test Drive

Carefully slide the paper clip from the balloon, and watch it float, as if by magic, across the table.

Alternate Model

For another version of the air car, glue the plastic spool onto the record so that the spool's hole and the record's hole line up. Inflate the balloon and clamp the neck with a paper clip. With the record on a flat surface, stretch the lip of the balloon over the spool and remove the paper clip. The record will glide across the surface. The added weight of the record prevents your air car from lifting as high, but it prvides added stability.

Why Airplanes Fly

You Will Need

- Hair dryer
- 8½ × 11-inch (21.25 × 27.5-cm) poster board
- 2 plastic drinking straws
- Kite string
- Hurricane lamp glass chimney
- ½ cup (120 ml) of puffed rice cereal
- 2 strips of stiff 1 × 3-inch (2.5 × 7.5-cm) cardboard
- Strip of thin 2 × 3-inch (5 × 7.5-cm) cardboard or poster board

Bernoulli Effect

How can a 20-ton (22–metric ton) airplane soar thousands of feet or kilometers above the earth as gracefully as a crane? It has to do with the **Bernoulli effect,** named after the scientist who discovered it.

The effect may be simply stated: As the *velocity* of a gas or liquid is increased, the *pressure* perpendicular to its direction of flow is decreased.

The top of an airplane wing curves so that air flowing over it speeds up. Fast-moving air means less pressure above the wing and more pressure below the wing, where the air moves more slowly. As a result, the wing is lifted from below and the airplane rises.

Flying Wing & Grounded Duck

For a clear demonstration of the Bernoulli effect, construct a flying wing and grounded duck. To make the wing, a simple airfoil, cut the poster board in half so that you have two strips 4¼ × 8½ inches (10.62 × 21.25 cm). Fold the first strip in half lengthwise. Starting from the fold, rub half of the strip against a tabletop until the poster board begins to curve. Then fold the curved half against the straight half, connecting the two halves with a little tape at the edges.

To make the grounded duck—actually, the control for this demonstration—take the second strip of poster board and fold it in half lengthwise. Fold it again ½ inch (1.25 cm) from the center fold you just made. Refold the strip along the second fold so that one edge of the strip protrudes. Fold ½ inch (1.25 cm) from the edge of this protruding strip. Tape the edges together to make a shallow box shape.

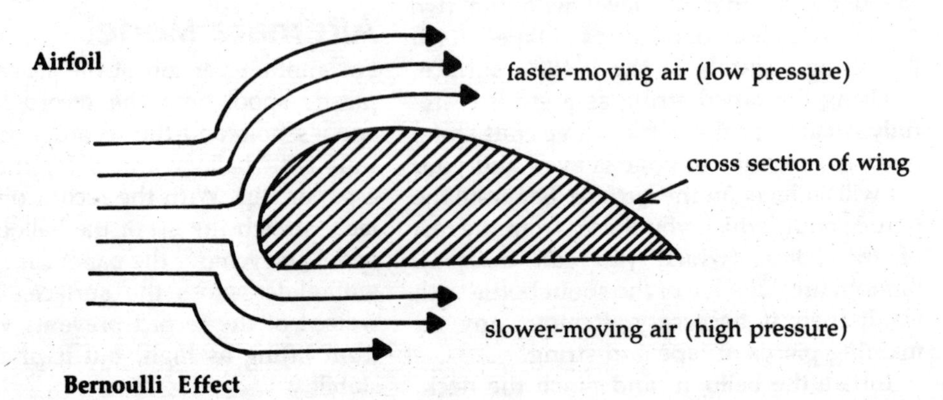

Bernoulli Effect

With the nail, punch holes in both wing and duck for inserting the straws. Cut both straws in half, and carefully poke them through the holes.

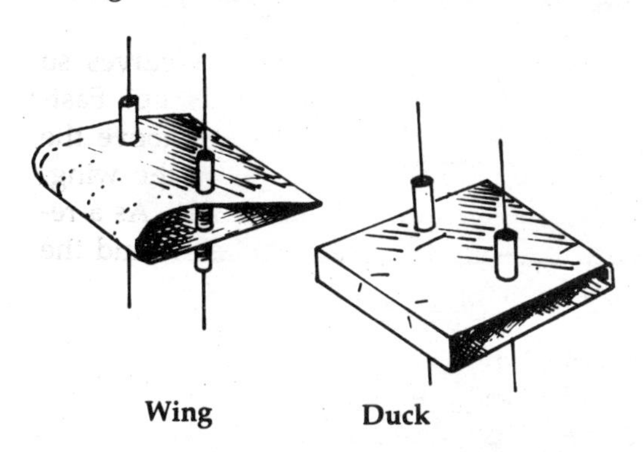

Wing Duck

To mount your wing and duck, use masking tape to attach four equal lengths of kite string to the underside of a table or ledge. Thread the string through the straws on the wing and duck. Fasten the other ends of the string to the floor with tape, making sure the string is taut.

Point a hair dryer towards the rounded end of the wing, and watch the wing rise, guided by the string runners. Do the same for the duck—it hardly moves at all.

Floating Rice

For another demonstration of the Bernoulli effect, draw a vertical line down the center of each cardboard strip. Cut halfway down the lines; then slide the edges of the strips together at the cuts so that both strips form a sturdy, cross-shaped support for the glass chimney.

Set the chimney on top of the support. Tape the poster board into a loop to create a fence around the chimney. Pour the half cup (120 ml) of puffed rice into the opening at the chimney's top.

Hold the hair dryer so that the stream of air blows across the top of the chimney. What happens? The air, blowing rapidly across the top of the chimney, allows atmospheric pressure to increase at the chimney's bottom so that the rice rises.

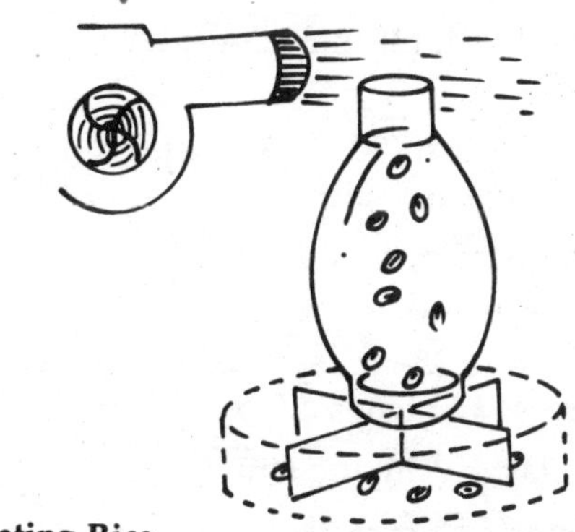

Floating Rice

The Bernoulli effect is also important in the design of hydrofoils and submarines. Understanding this principle aids the sailor, too. If he positions his sail correctly, he can move his boat forward *against* the wind.

Aerodynamics in a Wind Tunnel

You Will Need

- Small vacuum cleaner with hose attachment
- 12 × 24-inch (30 × 60-cm) plywood panel
- 8 × 12-inch (20 × 30-cm) plywood panel
- 12 × 16-inch (30 × 40-cm) piece of clear plexiglass
- 2 strips of plywood 5 × 24 inches (12.5 × 60 cm)
- ½-inch (1.25-cm) strip of plastic foam 5 × 12 inches (12.5 × 30 cm)
- 10 plastic drinking straws
- ¼-inch (.62-cm) diameter wooden dowel 2 inches (5 cm) long
- Hand drill with ¼-inch bit
- Coarse sandpaper
- Heavy-duty black trash bag
- Black cloth tape
- Epoxy glue
- Awl or knitting needle
- Piece of elastic
- Flat black paint
- Small, temperature-resistant bowl or can, less than 5 inches (12.5 cm) in diameter
- Plate to fit over bowl or can
- Dry ice
- Tongs
- 2 pot holders
- 4 wood blocks—3 cut into these shapes: equilateral triangle, rounded triangle, and teardrop *or* use blocks from a child's building set

Wind Tunnels

Wind tunnels provide engineers with valuable information. Everything from airplanes or cars to motorcycle helmets can be tested for good aerodynamic design in this simple apparatus. You can easily build one yourself to test basic shapes, cut from wooden blocks.

Tunnel Construction

Begin by standing the two 5 × 24-inch (12.5 × 60-cm) strips of plywood edgewise on a flat surface. Place them horizontally and parallel to each other, separated by 12 inches (30 cm). If they will not stand edgewise on their own, have someone hold them for you. Next, carefully lay the 12 × 24-inch (30 × 60-cm) plywood panel across the strips horizontally, aligning the top and bottom of the panel with the edges of the strips. Nail the panel to the strips and flip the structure over. Paint all exposed surfaces flat black and allow the paint to dry.

Assemble the vent piece from soda straws and plastic foam. Lay the plastic foam on a flat surface and divide it in half lengthwise with a line. With the knitting needle or awl, punch holes along the line, spacing them ½ inch (1.25 cm) apart. Twist the needle in the holes to widen them enough to contain the straws. Cut the soda straws into 1-inch (2.5-cm) pieces, and push them through the holes. They should protrude from just one side of the plastic foam.

Measure 8 inches (20 cm) from the left side of the structure, and insert the vent vertically with the protruding straws facing right. Make sure the vent sits evenly and at right angles to the structure's top and bottom sides (this is important for clear and straight trails of smoke). Apply a little glue to the edges of the plastic foam to secure it in place.

Next, lay the 8×12-inch (20 × 30-cm) plywood panel across the left side of the structure so that it fits neatly over the vent to create a box. Nail the panel securely in place, and paint it black also.

Construct the block stand by measuring 6 inches (15 cm) up the right side of the box and 8 inches (20 cm) towards the center. At that point drill a hole and widen it with sandpaper so that the ¼-inch (.62-cm) wide dowel fits snugly. Apply a little glue to the end of the dowel to keep it from slipping out. Drill a hole in the center of each block so that each fits over the dowel too, but can be easily removed and replaced. Attach the square block to the stand for your first test.

Finally, squeeze epoxy glue over the remaining exposed edges, and drop the plexiglass window into place. Allow glue and paint to dry overnight before the next step.

Stand your wind tunnel on its long side and check for warps and weak joints, making repairs if necessary. Attach a trash bag to the plexiglass side of the tunnel and fit the bag to the vacuum hose.

Place the open end of the trash bag around the mouth of the tunnel, fold in the excess, and wrap it tightly with cloth tape. Make a hole in the opposite side of the bag and insert a foot (30 cm) of vacuum hose. The hole allows the hose to fit snugly around it. Twist the excess bag around the hose, and tie it tightly with elastic or a rubber band so that no air can escape.

Blowing "Smoke"

In the final step you'll produce smoke and draw it through the tunnel. Watch the "smoke" collide with the block shape and observe the resistance patterns. Dry ice works nicely for the smoke, but you need to be especially careful with this dangerously cold material. *Never handle dry ice directly; use tongs for dropping the pieces into the temperature-resistant bowl.* With pot holders, lift the bowl into the hollow space at the left side of the tunnel, placing it close to the vent.

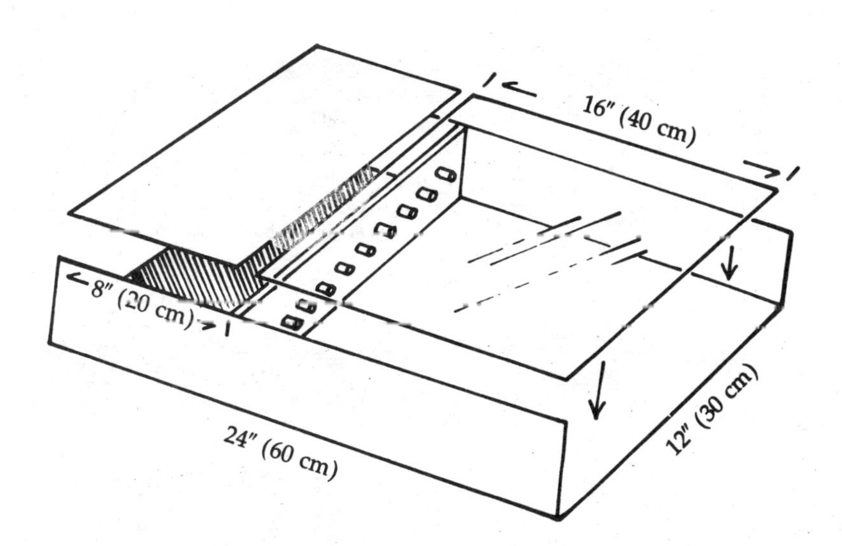

Wind Tunnel Assembly

Observing "Smoke" with a Little Assistance

Have an assistant pour water over the dry ice so that a thick, white smoke begins to form. Turn on the vacuum, and have your assistant pull the hose so that the bag stretches tightly. The suction of the vacuum will begin to pull long streams of smoke through the straws and across the empty area behind the plexiglass window. As the streams collide with each block, notice how they curl, break up, or disappear altogether. The amount of smoke disturbance around and behind an object indicates the object's *resistance* to smooth air flow and indicates poor aerodynamic design. Observe and take notes. Make a sketch of what you see.

Stress Test

Test the square first and notice how the smoke disintegrates completely as it collides head-on with the flat surface. A square, with its flat side facing the wind, makes a very poor aerodynamic shape. However, aerodynamics engineers have found ways to cut down the resistance of square objects. Trailers of large trucks become more aerodynamic by adding curved hoods that fit from the top of the driver's cab to the top of the trailer. Such designs are usually tested in a wind tunnel.

Turn off the vacuum and remove the hose from the trash bag. Reach through the hole in the tunnel and gently rotate the square 45 degrees. Reconnect the hose to the bag, and prepare the tunnel for operation.

Notice how much less resistance the block has when its corner faces the "wind." What conclusion can you draw from this? Notice how a considerable disturbance remains around the top and bottom sides of the block. What can you say about the aerodynamic qualities of a square?

Testing Other Shapes

Repeat the test with the triangle and rounded triangle, each time rotating them 45 degrees and comparing resistance patterns. What can you say in general about the aerodynamics of a triangle compared to a square? Record your observations and conclusions.

Finally, position the teardrop shape with its point facing the smoke. Turn it 45 degrees and compare the resistance. Does the teardrop appear more aerodynamic than the other shapes, even when turned? Make a list of familiar objects having this shape and how the shape might be useful to their operation.

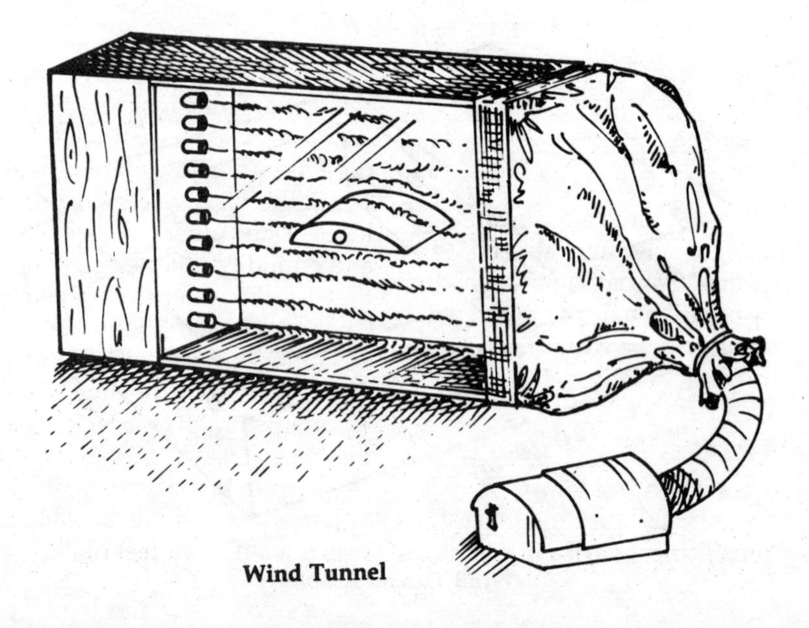

Wind Tunnel

Do Fans Really Cool You?

You Will Need

- 2 shallow, flat-bottomed pans, same size
- Warm water
- Eyedropper
- Small electric fan
- Clock

Dry-Air Relief

On hot days, everyone welcomes the breezy relief of an electric fan. But since a fan does not change the air temperature, how does it actually cool you?

A well-positioned fan certainly helps circulate and mix air, preventing hot air from settling on top and cool air from sinking below. But the main effect a fan has on cooling the body has to do with evaporating sweat from the skin's surface.

On hot days your skin perspires. If the air is very still, water vapor from perspiration saturates the air close to the skin. The humid layer of air acts almost like a layer of clothing, raising the skin's temperature. The breeze from a fan pushes this layer away, replacing it with a fresh, dryer supply of air which rapidly absorbs the perspiration and cools the surface of your skin.

A similar layer of humid air appears over lakes, swimming pools, and any body of fresh or salt water. Saturated air keeps the water from evaporating quickly. When wind replaces the saturated air with drier air, the water level drops noticeably.

Breeze Test

You can prove this with a simple demonstration. Be especially careful since this project uses an electrical device in combination with water. *Never place the water close enough to wet the fan.* Secure the fan firmly in place to keep it from tumbling over.

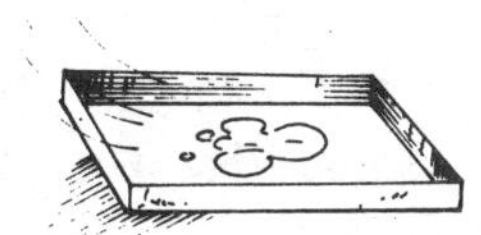

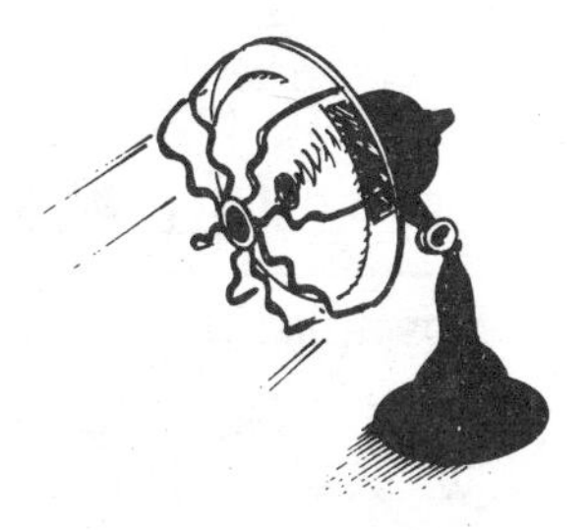

Place a fan 16 inches (40 cm) in front of the first pan. Switch it on and check to make sure a breeze blows directly across the bottom of the pan. If the fan oscillates (moves from side to side), turn off the oscillating switch.

Set the second pan far enough away from the first pan to protect it from the breeze, or place some kind of partition, such as a stack of books, between the two pans. When the setup looks good, turn off the fan, and place the clock conspicuously between the two pans.

Using the eyedropper, fill each pan with just enough warm water to barely cover the bottom. A graduated beaker will provide precise measurements, but if you carefully count the drops, the quantity of liquid in each pan will be equal.

Note the time just before you turn on the fan to begin the experiment. Then watch as the breeze-swept water begins to disappear while the still water remains at the same level. Depending on overall temperature and humidity conditions, it might take some time for the water to completely evaporate. Check at regular 15-minute intervals, then at shorter intervals when the water appears very low. Record the time it takes the water to completely disappear. Compare it with the time it takes the still water to disappear—usually a day or two.

If the pans were swimming pools, in which would you feel cooler?

Hot-Air Balloon

You Will Need

- Thin tissue paper
- Four ¼-inch (.62-cm) square balsa-wood strips 14 inches (35 cm)
- Four ¼-inch (.62-cm) square balsa-wood strips 3 inches (7.5 cm)
- Very thin iron wire
- Wad of absorbent cotton, about the size of an orange
- White paste
- Lightweight thread, cut into four 5-foot (150-cm) lengths
- Small fisherman's weight
- Light-colored poster board
- Masking tape
- Rubbing alcohol or Sterno paste
- Coffee can
- Tin snips
- Long kindling match

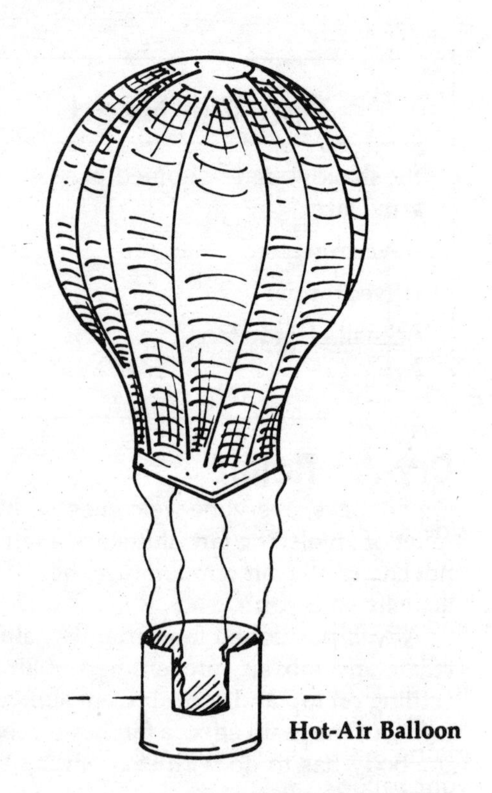

Hot-Air Balloon

Hot-Air Advantage

Hot air, lighter than cool air, rises. The French Montgolfier brothers amazed 18th century Parisians with a hot-air balloon that paved the way for more sophisticated balloon crafts using lighter-than-air gases instead of hot air.

Balloons have proved useful in the 20th century, from the great hydrogen airships of the 1930s to the ultra-buoyant weather balloons we use today.

Constructing a Paper Balloon

Reproduce the first pattern along the poster board's edge and cut it out. Turn the piece upside down and trace it on the opposite edge of poster board. Cut it out also. With masking tape, join the two pieces together along their straight edges. Make sure the sides are at a 45-degree angle from the pointed top. This completed piece functions as a pattern or template for tracing the balloon panel shape on the tissue.

In preparing the tissue, peel individual sheets away from each other so that you have the thinnest material possible. Carefully trace eight panels onto the tissue and cut them out. Place the first panel vertically in front of you, and apply paste along the right edge from the tip to the panel's flat

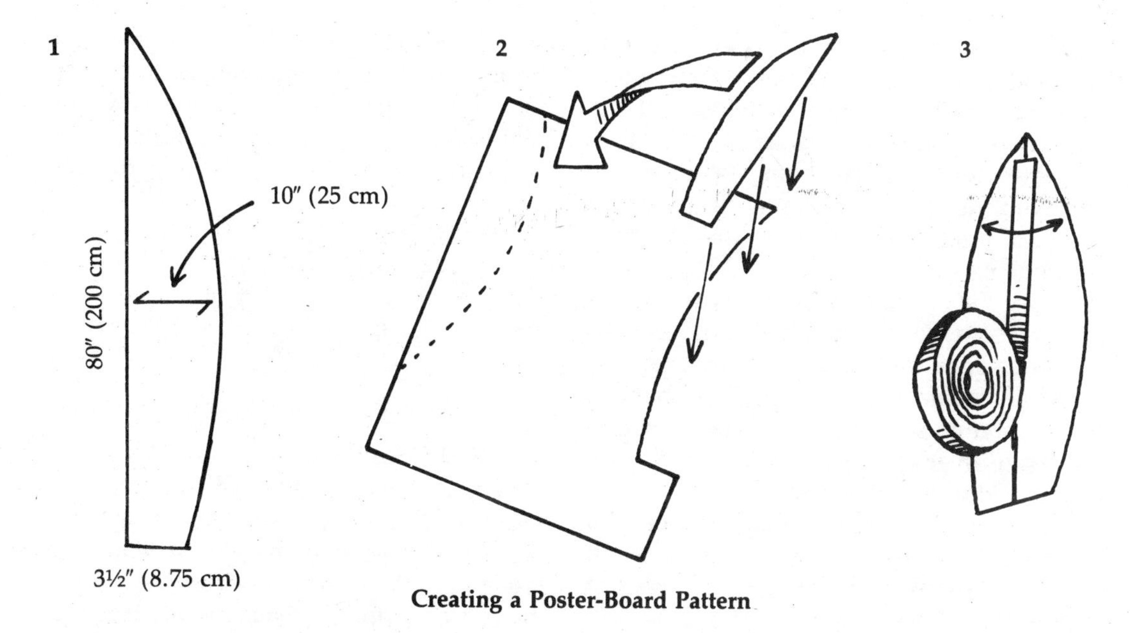

Creating a Poster-Board Pattern

side. Overlap the second panel with the first and press down against the glued edge. Now, glue the left edge of the second panel and overlap a third panel, pressing the glued edges together. In this way, continue to assemble panels 1 to 8 of your balloon accordion-style. When you've completed gluing the edge of panel 8, carefully unfold your balloon canopy so that panel 8 can be attached to panel 1, closing the envelope. Finish the balloon by trimming excess tissue at the top and sealing the top with another small piece of tissue.

Frame & Flame

The second part involves constructing the balsa-wood frame from which the alcohol-soaked cotton hangs. With paste, connect the four 20-inch (50-cm) strips of balsa wood together to form a square. Then paste the four 7-inch (17.5-cm) strips diagonally in the corners to reinforce the square. Use very little paste to attach the balsa-wood strips to each other since you want your craft to be very lightweight.

Next, carefully tie the thin iron wire from

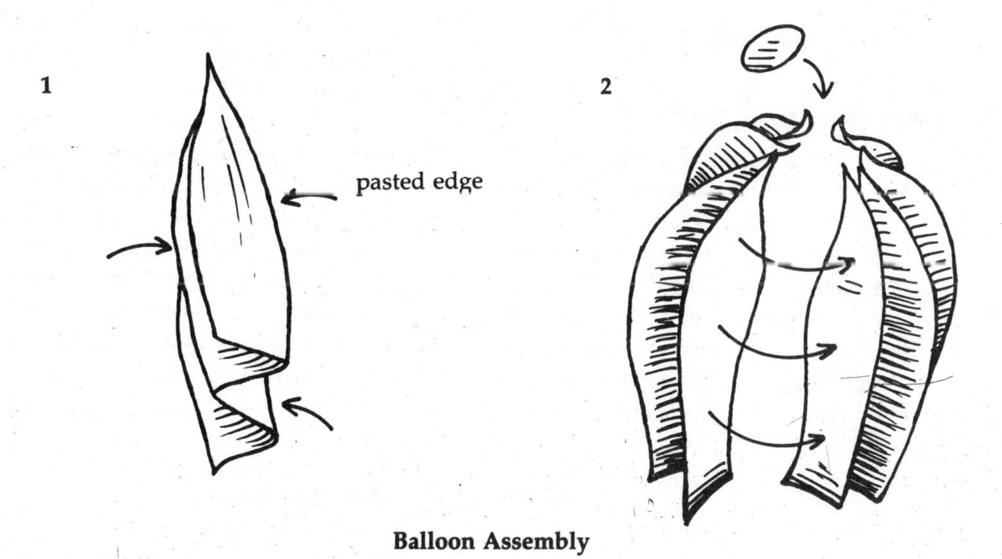

Balloon Assembly

corner to corner so that the wires intersect to form a cross in the center of the frame. Wrap a small amount of wire around the cotton wad, allowing enough wire to protrude to form a small hook. When we're ready to launch our balloon, we'll attach this hook to intersecting wires at the center of the frame.

Place the balloon on its side with its bottom facing you. Apply paste to the frame's outside edges. Holding the frame vertically, carefully maneuver it just inside the balloon's bottom so that about ¼ inch (.62 cm) of tissue overlaps the pasted edge of the frame. Trim off excess tissue.

Since the first flight of your balloon will be controlled, or captive, attach a 5-foot (150-cm) length of thread to each corner of the frame. These "anchoring threads" will be knotted together at opposite sides and attached to a small fisherman's weight.

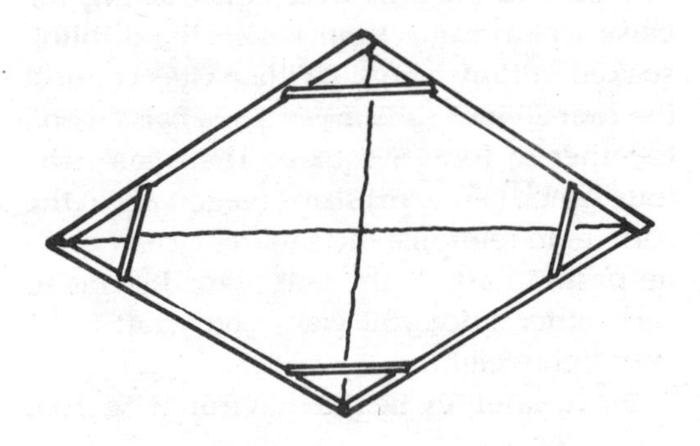

Balsa-Wood-and-Wire Frame

Preparing the Launch

Starting from the top of the coffee can, use the tin snips to carefully cut away a window about 3 × 5 inches (7.5 × 12.5 cm). Be careful. The cut-tin edges are very sharp. If possible, use a metal file to dull the edges.

Now, place the coffee can on the floor, open side up. Carefully stretch out the balloon envelope. Lift the fisherman's weight and attached threads and place them in the coffee can before lifting the entire craft onto the can. It should rest level without support. If this is not the case, have an assistant hold it upright until launching time.

The next part involves fueling the balloon, and since flammable materials are involved, be *extra cautious* and have an adult help you. Pour a very small amount of denatured alcohol into the cotton wad, just enough to moisten (not soak!) the cotton. Or, you can apply a small amount of Sterno jelly to the cotton wad, using a spoon. In any case, when you're through preparing the cotton, move the alcohol or Sterno to a safe place, far from your balloon.

Saturated Cotton Attachment

Maiden Flight

Holding the prepared cotton by its wire hook, carefully insert it through the window of the coffee can and attach it to the wires at the center of the balloon frame. Check to make sure that the cotton does not hang close to any of the anchoring threads. Spread the threads apart and pack them down in the coffee can if necessary.

Strike the kindling match and carefully ignite the cotton. *Make sure the match and the flame from the cotton do not come into contact with the balloon tissue.*

As hot air rises inside the envelope, the sides of the balloon will fill out and the craft will begin to lift from the coffee can. As this occurs, make sure the threads which anchor the balloon to the ground do not tangle and tip the frame. Allow your balloon to rise, drift gracefully at the end of its tether, and then slowly fall as the flame goes out and the air inside the envelope begins to cool.

Sturdiest Soap Bubble

You Will Need

- Liquid dishwashing detergent
- Drinking glass or small jar
- Shallow pan
- Plastic drinking straw
- Hot plate
- Small pot
- Pot holder
- Plastic bag
- Thermometer
- Stopwatch
- Air pump

Bubble Logic

Many scientists study bubbles because they provide information about physical laws and properties.

The bubble, a nearly perfect construction, demonstrates nature at its most efficient. The **sphere** contains the greatest volume within the smallest possible surface area. Tension at the surface is uniform for the entire bubble, and it creates an almost perfect spherical shape. This project concerns **surface tension,** important in a soap bubble's life, and the effects of temperature and various gases.

Bubble Test

Pour 1 tablespoon (15 ml) of dishwashing liquid into a glass and add 10 tablespoons (150 ml) of room-temperature water. Mix carefully to avoid making unnecessary small bubbles. Slowly pour the solution into the pan to cover the bottom with a shallow layer of liquid. Place the thermometer in a corner of the pan. Wait a few minutes; then record the temperature.

Have the stopwatch ready as you place one end of a plastic straw in the pan, and gently blow into the other end to draw out a bubble from the solution. Create a bubble that's large enough to study but small enough to fit freely and comfortably in the pan. Bubbles that collide with the sides of the pan may yield inaccurate data.

Snap your bubble free by plugging the straw's open end with your finger and twisting the straw sharply to the side. Start the stopwatch and wait for the bubble to burst, timing the interval. A bubble inflated by mouth contains mostly carbon dioxide.

Begin the bubble as before, but plug the straw before quickly attaching the air pump to it. Continue filling the bubble with pumped air until it reaches the approximate size of the first bubble. Snap it free and start the stopwatch.

Compare and record the amount of time it took the pump-filled and breath-filled bubbles to burst. Which bubble lived longer?

Cool the soap solution to about 55 degrees F (13 degrees C) by placing a few ice cubes in a plastic bag and tying the top so that no liquid leaks out. Place the bag in the pan and wait 15 minutes. Remove the bag and measure the temperature of the solution, making a note of it.

Create two bubbles as before—one breath-filled and the other pump-filled. Compare the time it takes each bubble to burst, and compare this with earlier data. Do bubbles drawn from a cold solution last longer than bubbles drawn from a room-temperature solution?

Finally, pour the solution out of the pan into the small pot. Place the pot on a hot plate and heat the solution to 120 degrees F (50 degrees C). Using the pot holder, carefully pour the solution back into the pan, and create two bubbles as before, recording data.

Best Bubble Lifespan

When you compare results you'll find: Bubbles filled with the carbon dioxide from human breath last longer than bubbles filled with air from a pump. Bubbles made from room temperature and cold solutions have the same lifespan, but bubbles made from heated solutions last nearly twice as long. *Warmth sustains surface tension* and prolongs the life of a soap bubble. Does this suggest why a breath-inflated bubble outlives a pump-filled one?

Freestanding Arch

You Will Need

- ¼-inch (.62-cm) plywood 9×16 inches (22.5 × 40 cm)
- ¼-inch (.62-cm) plywood 3×16 inches (7.5 × 40 cm)
- 2 pieces of ¼-inch (.62-cm) plywood 1 × 3 inches (2.5 × 7.5 cm)
- 2 small cabinet hinges with screws
- 2 feet (60 cm) of 1 × 1 wood
- 2 small nails or brads
- Cork
- Craft knife
- Small handsaw or band saw, if possible
- Carpenter's glue

Engineering Marvel

The strength of the arch makes it a versatile architectural form. Arches are put to work in bridge designs, monuments, and endless varieties of doorways, windows, and rooftops.

Unlike the triangle truss or square trestle, the arch contains no single point of structural weakness. Stress, either from above or below, is evenly distributed along the *curve* of the arch, and the interlocking pieces press even closer together. This makes it possible for even small arches, like those of the ancient Roman Colosseum and aqueducts, to support tremendous weight.

Making the Support Piece

Begin with the 3 × 16-inch (7.5 × 40-cm) piece of plywood placed over the two 1×3-inch (2.5 × 7.5-cm) plywood pieces, trestle-like. Make the smaller pieces flush with the edge of the larger piece. Attach each of the smaller pieces with two nails to create a kind of cradle. Turn the cradle flat side down and position it horizontally. At the top corner (facing you) of each small piece, attach a thumbtack.

Place the 9 × 16-inch (22.5 × 40-cm) piece horizontally above the cradle, and fasten the two pieces together at the edges with hinges. The pieces should pivot towards you. Turn the entire support on its edge and measure 4 inches (10 cm) along the edge of the larger piece. Hammer a nail there, allowing about ⅛ inch (.31 cm) of the nail-head to protrude. Turn the support on the

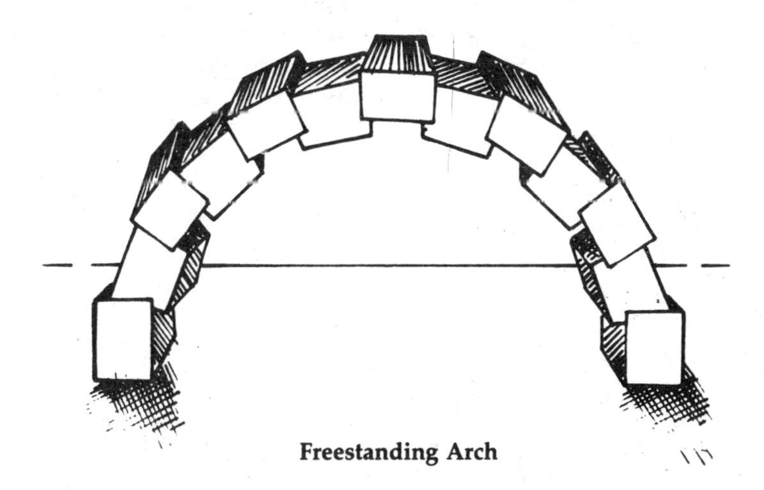

Freestanding Arch

opposite edge and repeat the procedure.

Using a craft knife, carefully cut the cork in half so that you have two semicircles with flat sides. Now measure 8 inches (20 cm) along the top edge of the cradle and mark it. Measure 6¼ inches (15.62 cm) in opposite directions from the mark, and glue the pieces of cork with flat sides facing in. The corks will act as anchors to prevent the arch's base from slipping outward when you turn it upright.

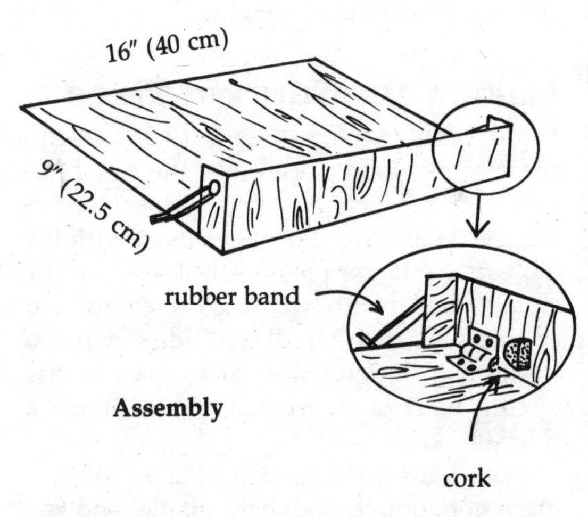

Assembly

Finally, fold up the smaller piece of wood so that it sits at a 90-degree angle to the larger piece. Attach rubber bands from the tacks to the nailheads on both sides of the support.

Arch Blocks

Construct the arch with thirteen 1 × 1½-inch (2.5 × 3.75-cm) interlocking blocks in two designs after the illustrated patterns, cut from the length of wood.

The second block design is trickier to cut out. Measure and draw the design on one of the blocks. Carefully cut out the pieces with

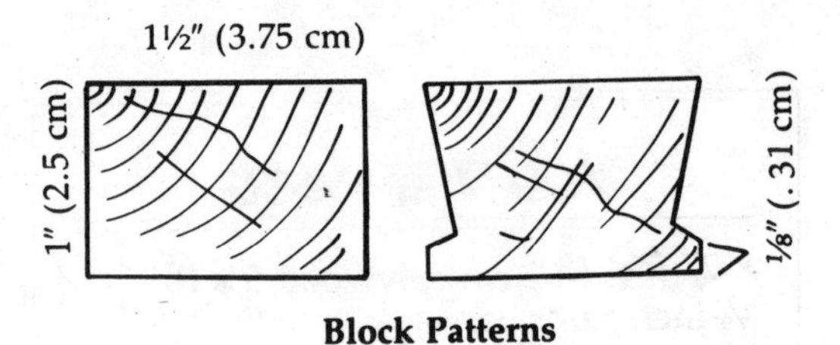

Block Patterns

a handsaw. Use this block as a guide for tracing the design onto five other blocks. You should make six cut blocks and seven plain blocks. Sand all blocks to remove rough and uneven edges.

Building the Arch

Begin assembling the arch with two plain blocks, placed vertically at opposite ends of the large plywood piece against the anchoring nails. Make sure the blocks do not rest against the hinges. Above the plain blocks, place two of the cut blocks, followed by two more of the plain blocks, until you have a completed arch lying sideways against the plywood.

To turn your completed arch upright, lift the larger piece of plywood by the top edge while carefully holding the smaller piece against it at 90 degrees. Keeping this position, slowly pivot the entire support until the arch sits upright. Make sure all blocks sit securely against each other; adjust them if necessary. Finally, remove the rubber bands and pivot the larger piece of plywood away. Your freestanding arch settles into itself as gravity pulls the blocks together.

Stress Test for Bridges

You Will Need

- 27 feet (325 inches or 8.3 m) of ¼-inch (.62-cm) balsa wood
- Craft knife
- Carpenter's glue
- Wooden block about 2 × 3 × ½ inches (5 × 7.5 × 1.25 cm)
- Eye-socket screw
- S hook
- 4 straight pins
- Shallow pan of water
- Small plastic bucket
- 2 tables of equal height
- Jar of pennies
- Scale (metric, if possible)

Ancient Bridges

Bridge-building dates back thousands of years. Ever since early humans first piled stones to make a footpath across a stream, constructing a road through air or over water has challenged the imagination.

Ancient bridges made of wood aided soldiers' movement across difficult terrain. Only a few stone footbridges from Roman times remain, but arched aqueducts designed to transport water hundreds of miles still work in many parts of Europe.

Iron Suspension Bridges

In the 19th century, iron began to replace wood as a bridge-building material. The strength and flexibility of iron cables and girders allowed engineers to experiment with new and bolder designs. Longer bridges became possible, and a new kind of bridge—suspended from trestles instead of

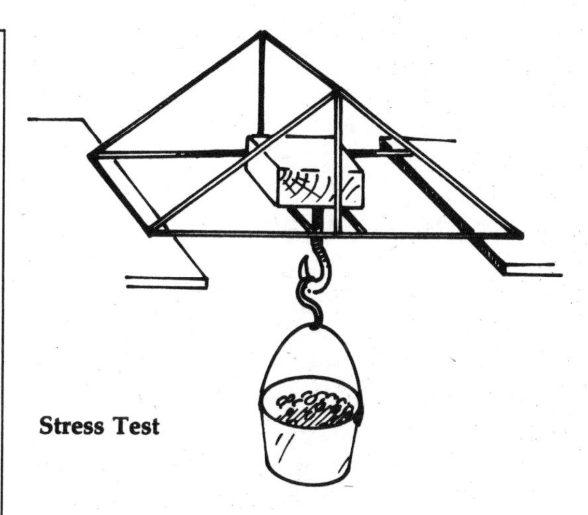

Stress Test

propped by a series of columns—made long spans across deep and treacherous waters less difficult.

Four Basic Bridge Designs

Good engineering remains the core of bridge building, however. We'll test and compare four basic designs.

Though easy to build, carefully assemble each bridge section by section, allowing the glue to dry before the next step. For all designs except the beam bridge, complete the side panels before connecting them with support pieces. Use books, if necessary, to keep your bridges upright until the glue dries.

The **beam bridge** consists of 14 pieces of balsa wood cut as shown. The **triangle bridge** uses 12 pieces, and the **double triangle truss bridge** uses 17 pieces, cut as shown. (See diagrams.)

The **arch bridge** requires a little more work since you need to soak the arch pieces until you can bend them easily. Fill a shallow pan with water, and submerge the two 16-inch (40-cm) strips of balsa wood, weighing them down with paper clips. While they soak, construct the side panels by gluing two 4½-inch (11.25-cm) vertical pieces to 12-

inch (30-cm) horizontal pieces at the center. Allow the glue to dry.

After a day or two, remove both arch pieces and test for flexibility. If they appear to bend easily, stretch them over the vertical piece of each panel, pinning each end to the ends of the horizontal piece. You should have a graceful bowed arch over each side panel. Allow the arches to dry before gluing them permanently in place.

Weights & Measures

To construct the weighing device, measure the length and width of the wood block, and bisect it along the length and width so that the intersecting lines reveal the exact center. At the center, screw in the eye socket until it is firmly attached to the block.

Prepare for the test by positioning two tables 10 inches (25 cm) apart, straddling the beam bridge evenly across the gap. Place the wood block, with eye socket facing down, between the bridge's support pieces. Now, carefully attach one end of the S hook to the eye socket, and the other end to the bucket handle. Be ready to record results.

Begin by adding pennies to the bucket, one at a time, counting as you go. Eventually, the weight of the pennies will cause the beam bridge to collapse, and the bucket will spill on the floor. Gather all the pennies, add them to the bucket, weigh the bucket, and record your data.

Repeat this procedure for the three other bridges, recording both the number of pennies and weight of the penny-filled bucket at the time of collapse. For a more accurate figure, you could first weigh the block and eye socket attachment alone, adding it to the bucket weight.

Notice the increasing superiority of design from beam bridge to triangle truss to double triangle truss and, finally, to arch. The arch supports twice as much weight as the triangle truss design, and nearly three times the weight of the beam bridge. Display your results in a chart.

Count the number and kinds of bridges you see the next time you travel. Why do you suppose the arch design outnumbers the rest?

Beam Bridge

Triangle Bridge

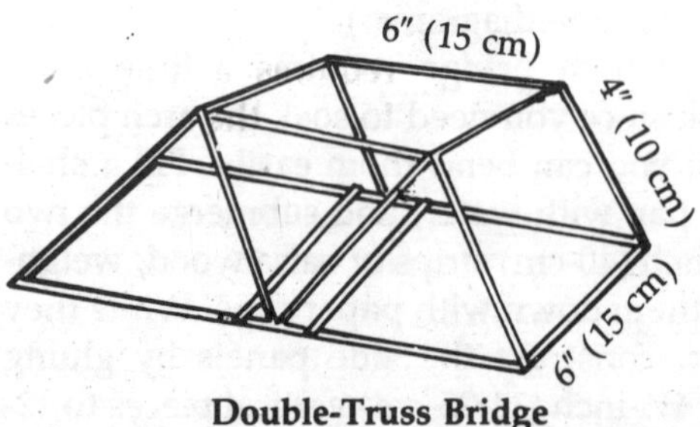

Double-Truss Bridge

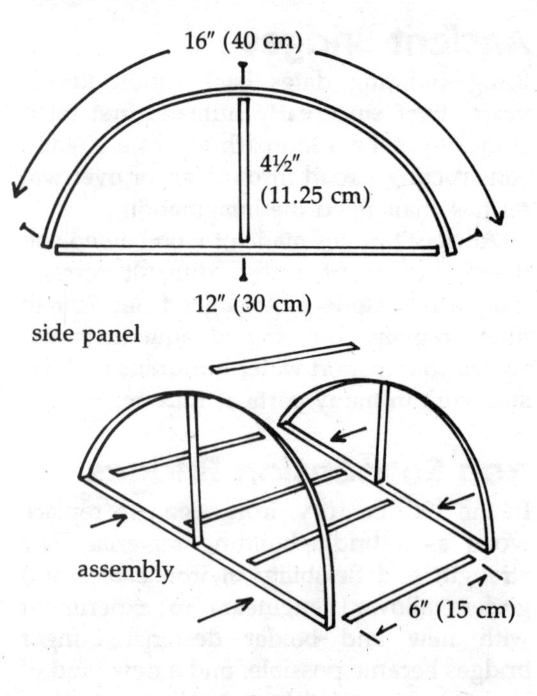

Arch Bridge

Amazing Life Forms

Leaf Shadow Prints
Smart Plants
Photos from Photosynthesis
Fern Life Cycle
Pomato
Spider-Web Collecting
Micro-Aquarium
Ant Farm
Ant Telegraph
Homemade Wormery
Frog Hibernation
Miniature Ecosystem

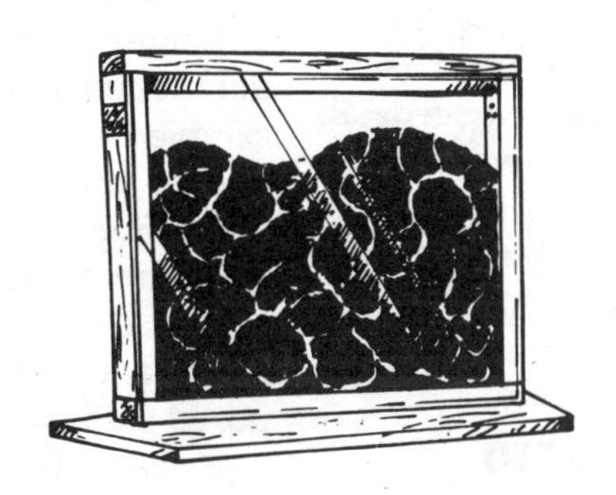

Leaf Shadow Prints

You Will Need

- Light-sensitive ozalid or diazo paper
- Bottle of ammonia
- Large mayonnaise (or similar) jar with cover
- Packet of fish tank gravel
- Black construction paper
- Photo-frame glass
- Leaf and plant samples

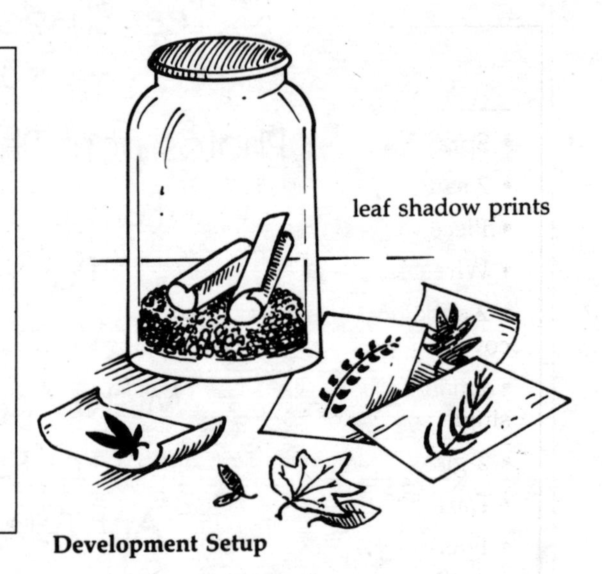

Development Setup

Picture This

You'll appreciate the complexity and beauty of plant forms when you see them in silhouette. With a few materials and a little patience, you can create an impressive collection of framable shadow pictures.

Choose plant samples with care. Ferns, small flowers, pine twigs, serrated (saw-tooth-edged) leaves, and winged seeds all have interesting shapes. If you want to collect samples now and preserve them later, place them between book pages to flatten them—the flatter the sample, the sharper the silhouette.

Homemade Darkroom

In a darkened room, cut the diazo paper into squares a little larger than the plant samples. Place the squares between several sheets of black construction paper to protect them from light.

To make a fixing jar, pour about ½ inch (1.25 cm) of ammonia into the mayonnaise jar, followed by 1 inch (2.5 cm) of aquarium gravel. Since ammonia can burn your lungs, be especially careful about inhaling it. Screw the cap on the jar and put it aside.

Remove a square of diazo paper from under the construction paper, and place a plant sample over it. Place paper and plant in the sun or under a strong light. Cover with photo-frame glass. Wait about 1 minute or until the paper surrounding the plant becomes colorless. Remove the glass and plant, and roll up the square with the exposed side facing in.

Keeping the paper rolled, drop it in the fixing jar and replace the cover. After about five minutes, the ammonia fumes will "fix" the shadow and make it permanent. It may take a little practice to get a clean shadow, but your effort will be worth it.

Smart Plants

You Will Need

- Spray bottle
- 2 saucers
- Piece of wire screen
- Wire cutter
- 2 shoe boxes, with and without cover
- Small potted potato plant with a single sprout
- 2 glass-covered photo frames
- Dark-colored felt or blotting paper
- Eyedropper
- Packet of mustard or radish seeds
- Lima bean seeds
- Shallow pie plate

Plant Tropisms

To find what they need, plants mysteriously know in which direction to grow. This type of growth pattern is called a **tropism.** The following project demonstrates the three basic kinds of plant tropisms—**hydrotropism,** towards water; **geotropism,** towards the pull of gravity; and **phototropism,** towards light.

Hydrotropism

You can easily demonstrate hydrotropism—a plant's growth towards water. Cut a 4-inch (10-cm) diameter hole from the bottom of the shoe box. With wire cutters, trim the screen so that it fits neatly inside the shoe box and over the hole. Cover the screen with about ½ inch (1.25 cm) of soil, and place a lima bean seed on the soil above the hole. Cover the seed with another ½ inch (1.25 cm) of soil, and rest the box so that the hole sits directly over the dry saucer.

Keep the soil moist, not wet. (A spray bottle comes in handy.) Avoid having excess water drip through the soil and into the saucer. If this occurs, remove the saucer and dry it thoroughly.

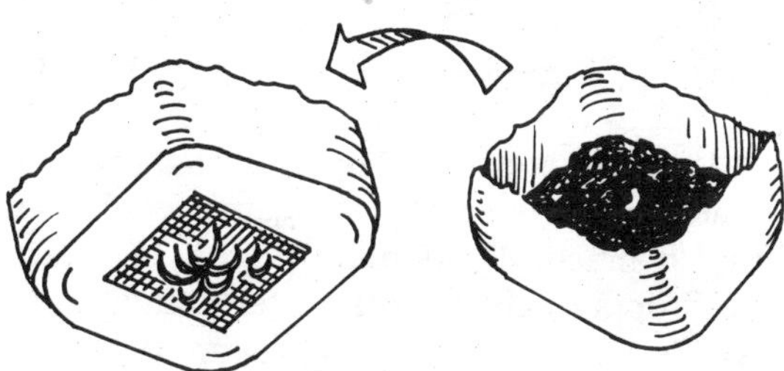

Hydrotropism

After about 10 days, lift the box from the saucer and look at the screen underneath. You should see small roots poking straight down, the abruptly turning sideways. The downward growth (geotropism) shows roots responding to the pull of gravity. But the sideways growth (hydrotropism) indicates their search for surface water, essential to survival. So important is this search that roots reverse themselves and grow in any direction where they detect moisture.

Repeat this procedure, but fill the saucer with water to about ¼ inch (.62 cm) from the screen. After 10 days, lift the box and notice how the roots poke straight down, detecting the water in the saucer.

Geotropism

You can also demonstrate geotropism—a plant's growth in response to gravity. Remove the panes of glass from two photo frames. Cut a piece of blotting paper or felt the same dimensions as the glass, and place it on one of the panes. Sprinkle a few mustard or radish seeds in the center of the felt, and cover the felt with a second piece of glass, like a sandwich. Loop two rubber bands horizontally at the top and bottom edges of the sandwich, and two more rubber bands vertically at the left and right edges.

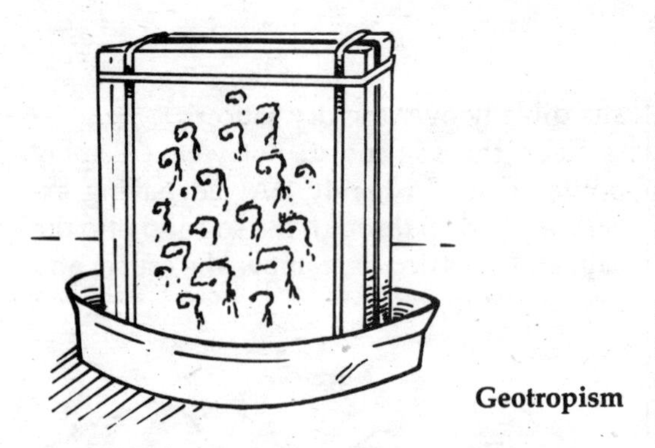

Geotropism

With its bottom edge in a shallow pie plate, lean the sandwich against books so that it practically stands upright. With an eyedropper, add enough water to the sandwich's top edge so that water seeps between the two panes of glass, wetting the seeds. Continue to water, and in about a week you'll see root hairs against the felt, growing straight down.

When the hairs become long enough, turn the sandwich on its side and continue watering for another week. Notice how the roots make a 45-degree change in direction. Now stand the sandwich upside down, and in a few days you will see the roots backtracking along their original direction. Finally, turn the sandwich onto its other edge and watch the roots make a complete loop in their travels.

Phototropism

Plants are experts at finding light. Remove the cover from the second shoe box and place it aside. With a craft knife, carefully cut out a 2-inch (5-cm) hole from the side of the box. Cut shirt cardboard into three strips as thick as the shoe box is deep and about three-fourths the width of the shoe box. Make a narrow fold along the strips at one end, and tape them equidistantly inside the shoe box, alternating sides. You should have three baffles and a small chamber at the shoe box's rear for the potato plant.

Place the potato plant in this chamber, and put the lid on the shoe box. Position the shoe box in a window with the hole facing the light. Open the box only to water the potato periodically and to observe plant growth. Eventually, you'll see the potato sprout snaking around the baffles towards the light.

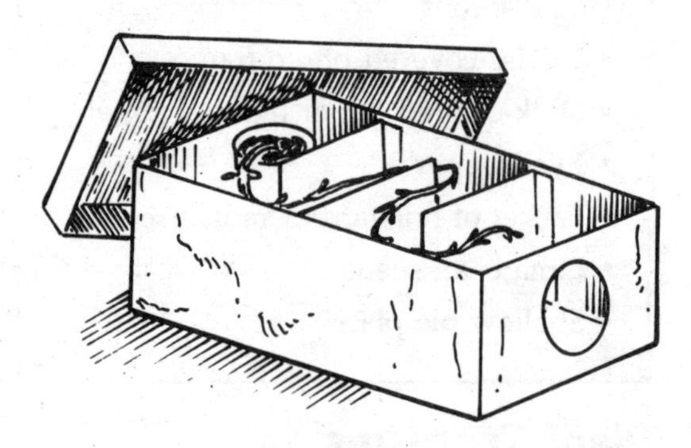

Phototropism

Tropism on Display

Display your three plant tropism samples side-by-side with an explanation of each. Remember, you need to plan well ahead for this project—your plants need to be fairly well developed to demonstrate these principles.

Photos from Photosynthesis

You Will Need

- Geranium or plant with large, thin leaves
- Small black-and-white photograph negative
- 3 × 5-inch (7.5 × 12.5-cm) index card
- Fluorescent light
- Hot plate
- Saucepan
- 2 small bowls
- Pyrex beaker large enough for plant leaf
- Tincture of iodine
- Isopropyl (rubbing) alcohol
- Craft knife
- Tongs
- Small towel
- Pot holder

How Plants Nourish Themselves

Plants nourish themselves through a process known as **photosynthesis. Chlorophyll,** which gives plants their green color, is the key to this amazing chemical process. Chlorophyll particles in plant cells use the sun's energy to transform carbon dioxide from air and water from soil into sugar and starch. With this unique experiment, you can see the starch produced in a plant leaf.

Photograph Negative Attached to Leaf

Preparing the Plant

Begin any project involving plant behavior well in advance. Allow at least four days before showing results.

First, cut the index card in half with scissors; then use a craft knife to cut a square in the middle of an index card half. The square should be a little smaller than the photo negative.

Fit the negative over the square and attach it with tape. Find a large outer leaf on the plant and staple the card to it. Two staples—one at the top and one at the bottom—of the card should be enough. Place the plant in a dark closet and wait 48 hours. Remove the plant and place it under a fluorescent lamp or a desk lamp for 26 hours.

Developing the Photo

Place about 1 inch (2.5 cm) of water in the saucepan, and bring it to a boil. Carefully remove the index card containing the negative from the leaf. Pry the staples free from the back to avoid tearing the leaf—your "photographic plate." Cut the leaf from the plant, and drop it in the pan of boiling water for about 30 seconds. Remove it with tongs. Pour about 1 inch (2.5 cm) of rubbing alcohol into the beaker, and place the leaf in the alcohol, spreading it out flat on the bottom.

Exercise caution in the steps that follow, and *always use the pot holder and tongs when handling hot materials.* Carefully place the beaker of alcohol in the pan of boiling water. If much of the water in the pan has boiled away, pour out what remains, and replenish the pan with 1 inch (2.5 cm) of fresh water, waiting for it to boil again before you continue. If there is too much water in the pan so that the beaker of alcohol seems unsteady and buoyant, pour out some water. *NEVER place the beaker of highly flammable alcohol directly on the hotplate!*

Allow the leaf to sit in the heated alcohol for about 9 minutes or until all the green coloring disappears. Then remove the leaf with the tongs and place it in the small bowl. Pour enough tincture of iodine over the leaf so that it is completely covered, and wait about 5 minutes. With the small towel, wipe the tongs clean of alcohol residue, then use the tongs to lift the leaf from the iodine, rinsing it in the second bowl of cold water.

Your photograph should stand out clearly. It's even suitable for framing!

Fern Life Cycle

You Will Need

- Large aquarium tank for the terrarium
- Soil and peat moss mixture
- Sand
- Digging spade
- Shallow flowerpot with dish

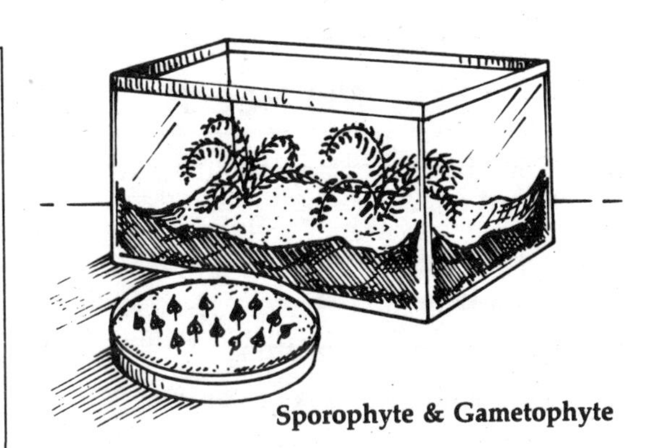

Sporophyte & Gametophyte

Sporophyte & Gametophyte

Those beautiful fiddlehead ferns growing in forest shadows tell only half the story behind this unusual plant species. Unlike the more sophisticated and less ancient plant forms, ferns do not spring whole from seeds. They grow into their mature asexual selves from an entirely different form of the plant. The fiddlehead or tree fern represents the more common *asexual*, **sporophyte,** half of this life cycle. The sporophyte produces *spores* which, in turn, generate the *sexual*, **gametophyte,** half of the cycle. This tiny gametophyte plant can then combine egg and sperm to create a new sporophyte, and the cycle continues.

Fern Results

By carefully cultivating both sporophyte and gametophyte, you'll see at a glance a fern's entire life cycle. Prepare the terrarium, filling it with about 4 inches (10 cm) of soil mixed with peat moss. Moisten it with a little water, and place the terrarium in a cool, shaded area.

Equipped with digging spade and newspaper, look for small ferns in shady wooded areas along the banks of ponds or streams. When you find one, dig carefully around it so that you can lift it out with roots intact. Wrap the fern in newspaper, and plant it in your terrarium, watering it frequently.

After a few weeks, check the fern's underside fronds for rows of tiny brown spore cases. If you see them, place the sheet of white paper underneath and gently tap the frond until a powdery brown dust appears. The dust is composed of thousands of spores.

Fill the shallow flowerpot with soil and peat moss, and cover this mixture with a layer of sand. Pour boiling water in the pot to kill potential molds that could destroy your plants. Allowing time for the pot to cool, place it in a dish filled with fresh water.

Carefully bend the white paper and sprinkle the spores into the pot. Cover the pot with cellophane wrap and continue to water the dish underneath. In several weeks you will see tiny, heart-shaped gametophyte plants, bearing no resemblance to the sporophyte fern. Soon, however, familiar fiddlehead shapes will emerge from the centers of these tiny plants and continue to grow until the gametophytes vanish. When large enough, remove the ferns from the shallow pot and transplant them in the terrarium.

Pomato

You Will Need

- Potted tomato plant, about 1 foot (30 cm) tall
- Potted potato plant, same size
- Craft knife
- Soft cotton cord or string
- Pliable grafting wax or paraffin

Grafting 1–2–3

Grafting projects require planning ahead. Allow at least 8 weeks from the time you graft the two plants to allow your "pomato" to flower and fruit.

You can grow the potato and tomato plants in separate pots or together in one large pot. When both are about 1 foot (30 cm) high, pull the main stems together. Where they touch, shave each stem with a craft knife just enough to expose the interior tubes. Tie the cut surfaces together with string; then, press the wax completely around the graft to protect it.

Allow about a week for the graft to take, checking for yellowing or withering on both plants. If the plants look healthy, cut off the top of the potato plant and the bottom of the tomato plant—turning your potato and tomato graft into a single "pomato."

Make several grafts and exhibit the best one in your display. When tomatoes appear above, you can be sure that potatoes grow below! Carefully remove a little soil to expose both to your viewers.

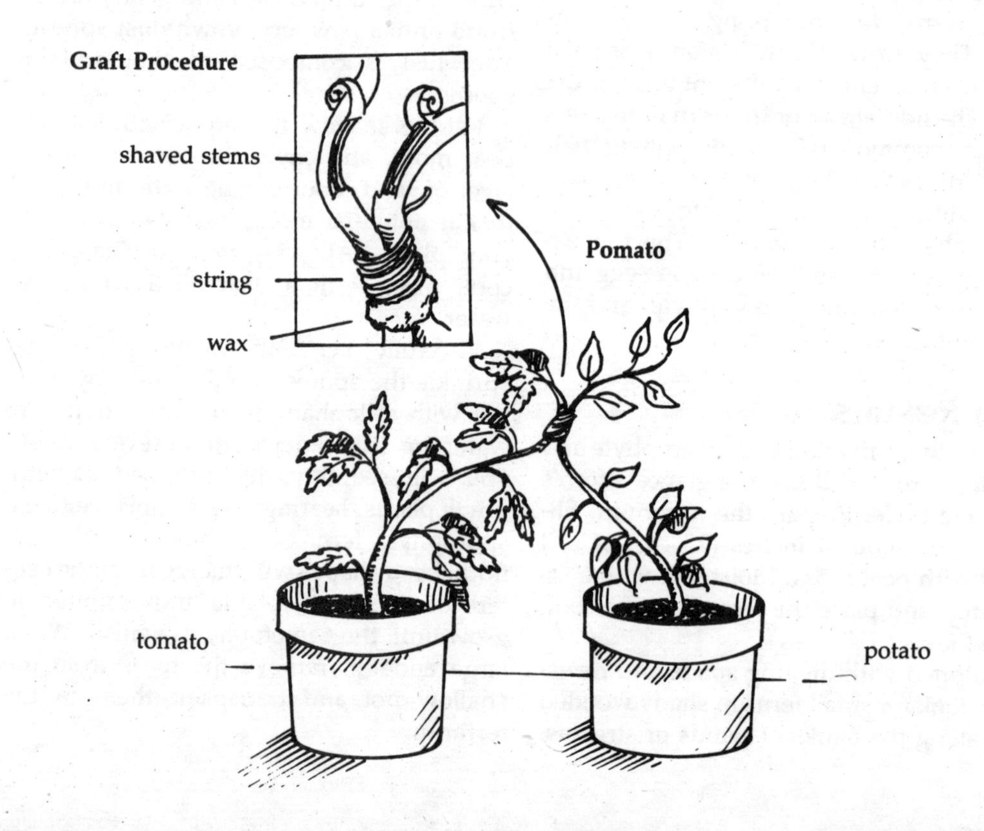

Spider-Web Collecting

You Will Need

- Sheets of black poster board, slightly larger than the spider webs
- Spray glue
- Insect repellent
- Sheet of clear, adhesive plastic
- Field guide for spiders in your area

Webs They Weave

Scientists have long puzzled over the strangeness of spiders and marvelled at the intricacy of their handiwork—the spider web. One thing is for certain: They're expert builders. In a recent space shuttle experiment, spiders performed nearly perfect feats of web-building in a weightless environment.

Noiseless Construction

An early morning spring stroll through the garden should reveal at least one excellent spider web specimen. You can easily accumulate a stunning collection of them with a few materials. Most spider webs are one of three designs—the cob web, triangle web, and orb web.

The most common of these, the **orb web,** displays the craft of the ordinary garden spider. Cob webs can have diameters well over 5 feet (150 cm). The **triangle web** is also very common but a bit smaller.

If you find an **orb web,** *however, beware! It's the type of design the black widow spider prefers. You should be cautious when approaching any web. Several species of spider other than the black widow also contain venom and their bites may be dangerous.*

Buy a reliable field guide to insect life in your area, and take it with you when you go out to collect. Familiarize yourself with poisonous species. You can also gently spray the web and surrounding area with insect repellent before you attempt to remove it.

When you find a web you like, spray glue along the edges of the poster board, keeping the inside glue-free. Position the poster board behind the web, and slowly move it forward until the edges of the web stick to the glue-covered edges of the board. Avoid moving the poster board any more or you might tear the web. With scissors, carefully cut around the web, severing supporting strands.

To preserve the web permanently, remove the backing from the sheet of plastic adhesive and carefully press it, sticky side down, over the poster board. Do this in one

Spider Web Varieties

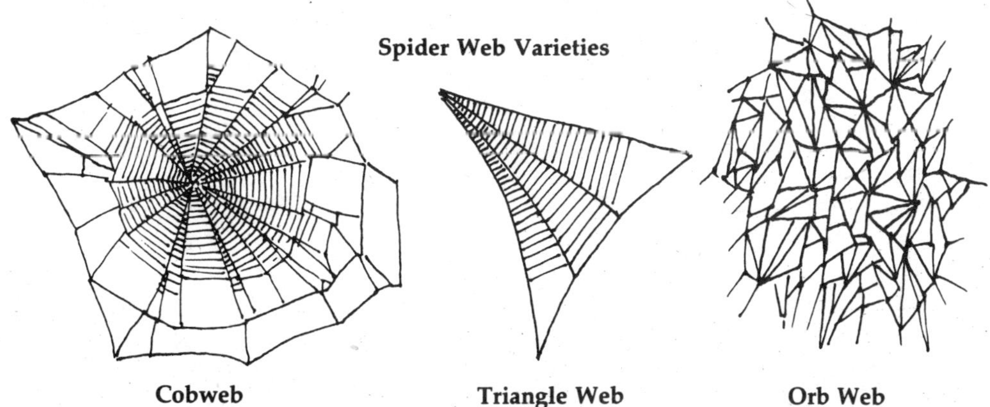

steady step; pulling back the adhesive and attempting to reposition it will destroy the web. Trim off extra plastic with scissors.

Mount your spider web specimens side by side in a display, noting the time, place, location, and date each was found. If you can identify the spider, note that as well. You will be amazed at the beauty and variety of webs in your collection.

Micro-Aquarium

You Will Need

- Filmstrip projector that displays lantern slides (check your school)
- Two 2 × 2-inch (5 × 5-cm) glass slides
- Long wooden matchstick or very thin strip of balsa wood
- Epoxy glue
- Craft knife
- Eyedropper
- Waterproof aquarium tape

Micro-Aquarium & Slide Projector Setup

Monstrous Projection

Imagine a 5-inch (12.5-cm) long water flea twitching across a projector screen, or a monster-size hydra swallowing bits of brightly colored algae. All this is possible with a miniature aquarium adapted for filmstrip-projector viewing.

Most schools have a standard filmstrip projector with clip mounts for slides. A large white sheet will work for the screen; curtain off both sides of the sheet to make the area dark.

To construct a micro-aquarium, place two glass microscope slides on a flat surface. Cut three 2-inch (5-cm) pieces of wooden matchstick, and with a little epoxy glue, attach one piece to the bottom of a slide. Attach the remaining two pieces to the sides of the slide and allow the glue to dry.

Trim the excess matchstick from the top of the slide with a craft knife. Apply a little glue along the edges of the matchsticks, and gently press the second glass slide over them. Allow the glue to dry. Finally, fold a thin piece of aquarium tape over all three sides to ensure that it's water-tight. You now have a miniature aquarium the width of a wooden matchstick.

Collecting aquarium specimens may seem challenging, but **microscopic organisms** abound in small lakes or pools of stagnant water. If you have a flowerpot with standing water in its dish, collect some with an eyedropper, and add it to your micro-aquarium.

Carefully slip the aquarium, open side up, between the clips of the filmstrip projector. Turn on the projector and focus. The light and heat from the projector bulb causes the tiny creatures to move rapidly. Their variety will amaze you. Add a little salt to the top of the aquarium and watch the effect it has on the creatures.

Ant Farm

You Will Need

- 2 glass-covered photograph frames 9 × 12 inches (22.5 × 30 cm)
- 50-inch (125-cm) strip of wood, 1 inch (2.5 cm) square
- Fine wire screen, 1 inch (2.5 cm) square
- Epoxy glue
- Carpenter's glue
- 1½-inch (3.25-cm) narrow-gauge nails
- 5 × 14-inch (12.5 × 35-cm) plywood
- Large white poster board
- Digging spade
- Bucket
- 2 small jars with lids
- Eyedropper
- Soil
- Cotton-tipped swab
- Honey
- Tweezers

Insect Colonies

Ants, bees, wasps, and other **colony insects** live in an elaborate and mysterious society. You can closely examine the inner workings of such a society by building a **formicarium**, or ant farm. This one is easy to construct and will provide hours of fascination.

Ant Farm Construction

Begin with the frame. Cut the strip of wood into five sections: two 12-inch (30-cm) sections, two 8-inch (20-cm) sections, and one 10-inch (25-cm) section. Apply carpenter's glue to one end of both 8-inch (20-cm) sections. Straddle one of the 12-inch (30-cm)

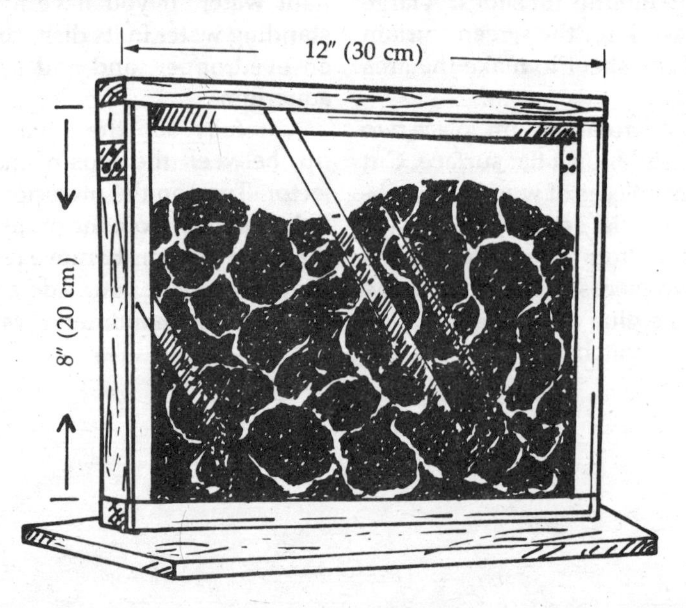

Ant Farm

sections between the glued ends, and hammer it in with nails. Glue the 10-inch (25-cm) section to the other 12-inch (30-cm) section, centering it carefully so that you have 1 inch (2.5 cm) clear at both ends. Allow the frame to dry overnight.

Remove the glass from photo frames and set the glass aside. Lay the frame flat, and apply a little epoxy glue around the edges of the sides and bottom. Carefully press one piece of glass against the glued edges so that it fits neatly over the frame. Wipe off any glue that seeps out under the glass. Allow this to dry about ½ hour; then, turn the frame over to attach the second piece of glass the same way. Allow the completed frame to dry overnight. Finally, turn the frame on its side and drill three small holes in a triangular formation about 2½ inches (6.25 cm) from the top edge. Place the piece of wire screen over the holes, attaching it with a little epoxy glue at the corners.

To construct the base, draw a horizontal line dividing the piece of plywood into two equal halves. Apply carpenter's glue to the line, leaving 1 inch (2.5 cm) dry at both ends. Center and place the frame over this line. Allow several hours for the glue to dry. Make sure the 12-inch (30-cm) lid fits neatly into the opening at the frame's top, using sandpaper, if necessary.

Recruiting for the Ant Colony

To populate your formicarium, pack two jars, a digging spade, and white poster board, and look for a dry, sandy field. Either small black or small red ants will do, but do *not* mix them! Also, use gloves and tweezers when handling red ants because some species bite. You can often find entrances to ant colonies under flat rocks.

Collect about 100 ants in one of the bottles, then gently dig down deeper to locate the **queen.** If you see ants scurrying in all directions carrying **larvae,** the queen is close by. Place the white piece of poster board flat on the ground, and gently deposit each new clump of soil, carefully breaking it up. Soon you'll see the queen—larger and paler than the other ants. Carefully lift her and place her in the second jar. Finally, fill the bucket with soil from the area.

Coax the ants into entering your formicarium by introducing the queen. But first, take the soil from the bucket and fill the frame to ¼ inch (.62 cm) from the lowest air hole. Lift the queen from her bottle and drop her through the frame's top. Then, carefully shake the rest of the ants through the top. Before placing the lid on the frame, dip one end of a cotton-tipped swab into honey and stick it in the soil.

Order Restored

For a while, your ants will race around in a panic. Cover your formicarium with an opaque paper bag, and leave it for about 12 hours. When you unveil your formicarium, you'll see the ants busily tunnelling. After a few days, an amazing and complicated network of tunnels and chambers begin to emerge. Identify some of these tunnels and guess what special purpose they might have. Draw a map of the tunnel and chamber system and place it next to the formicarium.

Moisten the soil periodically with the eyedropper. Feed the ants honey on a weekly basis, and your colony will thrive. Try placing different insects or bits of foreign material into the formicarium, and observe how the ants respond.

Ant Telegraph

You Will Need

- Formicarium (see Ant Farm)
- 2 × 2-foot (60 × 60-cm) shallow box
- Cellophane to cover
- White tempera paint
- Paintbrush
- Sugary fruit
- Cotton Swab
- Desk lamp

Insect Chemistry

Many species of insect possess amazing methods of long-distance communication. The moth uses feathery antennae—called **chemoreceptors**—to detect the scent of a female within a radius of 10 miles (16 km).

Ants also have an elaborate scent-based communication system due to substances produced in their bodies, **pheromones.** Signals sent with pheromones can be simple or complicated. With powerful pheromones, a single ant can telegraph important news to practically an entire colony.

The familiar sight of ants walking single file to a food source, then back again, is based on a **pheromone trail.** The first ant to find the food returns to the colony, depositing pheromones. Several ants then follow it to the food site, leaving a stronger pheromone trail behind for the rest of the colony. Interestingly enough, if the original ant strays, or meanders over a pebble or twig, other ants dutifully follow, even when a more direct route would seem more efficient.

Chemists have successfully synthesized some simple pheromones with amusing results. When one of these is placed on a pinpoint and drawn into a circle, the ants walk around and around in a futile search for food.

Ant Trail

You can demonstrate a simple pheromone trail by opening the top of your formicarium and placing a piece of sugary fruit about 2 feet (60 cm) away. Make your ants very hungry by not replenishing their honey for about 1 week. The best way to keep the ant activity confined and clearly visible to onlookers is to place both the formicarium and fruit within a shallow box, the inside of which you have painted white. Cover the box with a sheet of cellophane and place the lamp overhead for illumination.

In only about an hour, you'll begin to see a sparse line of ants making their way to and from the fruit. After several hours, two thick and busy lines of scurrying ants should be visible. With the ants busy, you can prepare the second demonstration.

Enemy Assault

One of the more complex pheromone signals warns of *danger* from an outside source. This pheromone is usually produced by just one or several ants, but it quickly spreads to others and mobilizes an entire colony. As pheromone molecules diffuse outward from the agitated ant, other ants within a 3-inch (7.5-cm) radius of the danger flee, while those outside that radius move in to attack.

Remove the cellophane covering the box. With the cotton swab, carefully poke one or several ants in the line. Quickly remove the swab and observe what happens. The line will begin to disintegrate as ants flee from the site of the disturbance. Watch along the perimeter of this activity as many more ants assemble and begin to close in on the danger—an impressive display of force. After a while, food lines continue as before. But you can repeat this dramatic little demonstration for hours or until the sugary fruit runs out.

Homemade Wormery

You Will Need

- Aquarium tank or large mayonnaise jar
- Thick cloth for covering
- Soil
- Leaf mulch
- Sand
- Cornstarch
- Stale vegetable foliage
- Earthworms
- Coffee can

Earth Moss

Of great benefit to farmers and flower growers, humble earthworms enrich soil by continually mixing and loosening it. They burrow tunnels that allow *oxygen* to enter soil, aiding the healthy growth of roots.

Worm Observatory

You can construct a wormery to observe slow-moving but busy earthworms at work. You'll need to prepare leaf mulch ahead of time. Making a usable mulch will take about a month. If making your own seems too much trouble, buy a commercial brand in any gardening store. Worms, however, appear to prefer homemade stuff.

To make mulch, punch as many holes around the sides of the coffee can as possible without denting it. Place dead leaves, vegetable and fruit peels, grass cuttings, cornstalks, or other organic materials in the can. If possible, also add wood ashes from a fireplace. Wet the contents and allow your mulch to stand for a week. Then turn over the compost with a spoon so that new material comes to the top about once a week.

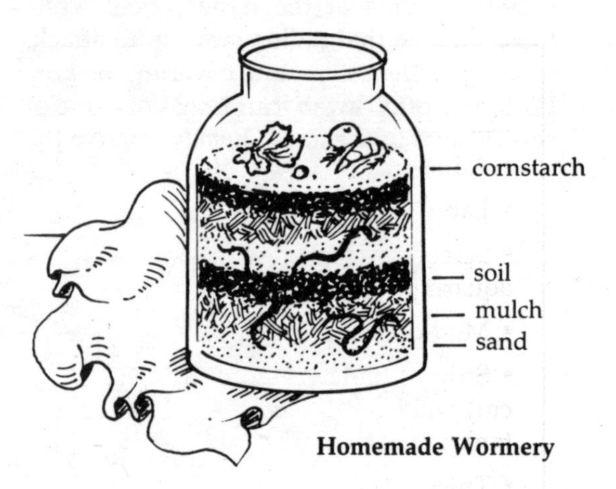

Homemade Wormery

Keep the mulch wet to aid decomposition.

Construct the wormery by layering sand, mulch, and soil, repeating the layers until you reach 4 inches (10 cm) from the jar's top. Make layers clearly visible outside of the jar, patting them down and levelling them with a spoon, if necessary. On top of the last layer, sprinkle a thin coating of cornstarch and cover that up with stale vegetable foliage. Keep the jar top open for air and pour in some water. Finally, add 4 or 5 worms. Look for them in your garden, on a garden path or sidewalk early on a wet morning, or after a good rain. Afraid of drowning, worms often come to the surface when it's raining.

Drape the jar with a thick cloth. Avoid letting the worms detect light since they will not burrow against the jar's sides where you can see them. Lift the cloth twice a week only to add more vegetable foliage.

After two weeks, uncover the jar, and record your observations. The layers of soil, sand, and mulch, once distinct, combine in crisscross patterns. Traces of white cornstarch appear everywhere—from top to bottom. And, because of the earthworms' tunnelling and loosening activity, there appears to be more soil in the jar than before. Rotate the jar carefully, looking for worms at work.

Frog Hibernation

You Will Need

- Adult frog
- Large aquarium
- Screen or chicken wire to fit over aquarium
- Mud containing clay
- Strip of tin lawn fence 1 inch (2.5 cm) wide and longer than aquarium tank width
- Thick cloth to drape over aquarium
- Lamp with infra-red bulb
- Ice cubes

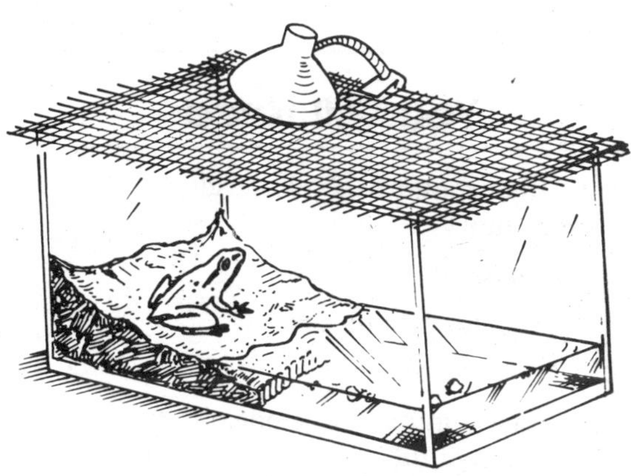

Frog Hibernation Setup

Long Winter's Sleep

When food is scarce and conditions harsh, many animals spend the winter sleeping. This winter nap, called **hibernation,** conserves the animal's energy and keeps food available for animals that do not hibernate.

Tricking a Frog

By creating a semiaquatic environment in an aquarium and adjusting its temperature, you can actually trick a frog into hibernating. The temperature change will not harm the frog.

Insert the strip of lawn fence in the aquarium and bend the edges against the aquarium's sides. Position the fence a little off-center to create two sections—a larger one for a mud bank and another for a pool. Fill the section behind the fence with mud; build mud high against the tank's left wall, and taper it down to the fence. Fill the other section with enough room-temperature water to creep over the mud bank.

Place the frog on the mud bank and cover the top of the tank with the screen. Allow the frog to relax a few minutes in its new home before proceeding.

Position the lamp with an infra-red bulb over the tank's top so that the light shines down evenly on the frog. Observe the frog as it grows more animated—moving about, rapidly pulsing its throat, possibly even going for a swim. Record your observations.

Switch off the lamp and drop a dozen ice cubes into the tank, both in the water and on the mud bank. Cover the tank for a few hours. When you undrape the tank, notice how the frog's breathing and movement have become slow and sluggish. If you wait long enough while continuing to add ice cubes, the frog will burrow into the mud bank and stop moving altogether, that is, hibernate.

Frogs on Display

For an effective demonstration, have two frogs on display—one active and one hibernating.

Miniature Ecosystem

You Will Need

- Large round fishbowl or globe terrarium
- Fine sand
- Pebbles, various sizes
- Fresh water
- Aquarium plants—elodea and myriophyllum
- Pond snails
- *Lemna minor* (a tiny shelled aquarium animal)
- Caddis gnats (found around ponds or lakes)
- Old nylon stocking
- Old white cotton pillowcase
- ¼-inch (.62-cm) wooden dowel, about 3 feet (90 cm) long

Crystal Bowl

We can demonstrate the complex relationship between plants and animals in a pond. If judiciously stocked with appropriate life forms, your experimental pond will require little or no maintenance for months and will become a self-regulating miniature ecosystem.

Pond, Plants, Populace

Place about 3 inches (7.5 cm) of fine sand in the bottom of the bowl, and arrange a few pebbles artistically. If you want some whimsy, add small ceramic trees and other structures. Fill the bowl with water more than halfway, and add the plants. Elodea and myriophyllum—used in combination or separately—provide oxygen and purify the water. Stock the bowl with about three or four small plants, rather than fewer large ones.

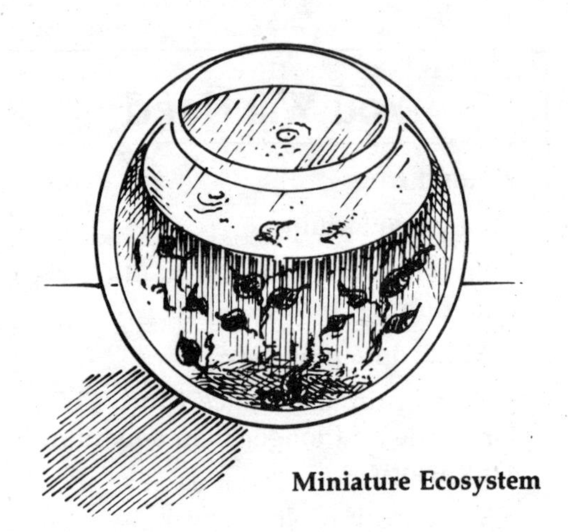

Miniature Ecosystem

To populate your bowl, find small snails and *Lemna minor* at aquarium shops, since they are often used to clean the insides of large tanks. You must capture the caddis gnat near the banks of streams, lakes, or ponds, using a homemade net.

Natty Gnat Net

To construct your net, place a pillowcase on a flat surface so that the open side faces you. Slit the pillowcase across the left side and top, and open it like a book. Fold the bottom edge up 1 inch (2.5 cm), and sew it so that you have a tube of material wide enough to poke a straightened wire hanger through.

After threading the hanger, bend it into a loop (if the hanger is short, bunch up excess material at the end) and twist the ends of the loop together to form a 1-inch (2.5-cm) piece that sticks straight out.

Carefully drill a hole into the dowel's end, and insert this piece. The fit should be tight. If the hanger moves inside the dowel, wrap tape around the twisted end before re-inserting it.

Swarms of caddis gnats populate freshwater shores, particularly towards evening. A few easy swings of your net should capture more than enough. Add them to the bowl, and cover the top with a piece of nylon stocking, attaching it with a rubber band or tape.

Ecosystem Observations

Keep a log as you observe your ecosystem from week to week. The caddis gnat lays eggs in the water, which sink to incubate in the sand at the bottom. Soon, tiny aqueous larvae appear. Those that survive become adult flies, eventually breaking from the water and forming swarms above it.

Display your ecosystem beside the caddis net with perhaps a few snapshots detailing stages of assembly. Include your log, updating it to note any clouding, discoloration, odor of the water, and any changes in the animals or plants.

[illegible] appears in the [illegible] ly [illegible]
in the [illegible]. Those that [illegible] to ne[illegible]
adult [illegible] eventually [illegible] from the [illegible]
and [illegible] around.
[illegible] your [illegible] the [illegible]
[illegible] will [illegible] questions [illegible]
[illegible] Include [illegible]
[illegible] [illegible] of the water, [illegible] in the
[illegible]

[illegible] into them [illegible] to the [illegible] of the
pond and cover the top with a piece of [illegible]
[illegible] stocking, attaching it with rubber band
[illegible]

Ecosystem Observations

[illegible] a log [illegible] observe your ecosystem
[illegible] week [illegible]. The [illegible] that [illegible]
[illegible] the water [illegible] which [illegible] includes [illegible]

Creative Chemistry

Homegrown Crystals

Solutions, Suspensions & Colloids

Adsorption Chromatography

Vegetable & Fruit Dyes

Creepy Corsage

pH of Cosmetics & Cleaners

Detecting Vitamin C in Fruits & Vegetables

Testing for Bacteria in Milk

Decomposition of Natural Plastic

Homegrown Crystals

You Will Need

- Hot plate
- Pot holder
- Graduated beaker (in cubic centimeters)
- Scale for measuring grams
- Thermometer
- Distilled water
- Spoon or stirrer
- Tweezers
- 6 medium-size mayonnaise jars with covers
- Small drinking glass
- 3 shirt cardboards
- Potassium sodium tartrate (Rochelle salt)
- Potassium aluminum sulfate (alum or potash alum)
- Potassium chromium sulfate (chrome alum)
- Potassium ferricyanide (red prussiate of potash)
- Copper acetate monohydrate
- Glacial acetic acid
- Powdered calcium oxide

Crystal-Clear Reflection

Chemists study **matter**—anything in the universe that has weight and takes up space. Chemists also study the way matter changes and combines with other matter.

The three states of matter—solid, liquid, and gas—each has its own properties and peculiarities. **Gases** consist of widely spaced and randomly moving atoms, the combined collisions of which create pressure. **Liquids** consist of more compressed atoms that move around freely. In **solids** these compressed atoms exist in relatively fixed positions. That is, they move by vibrating in place but cannot pass each other.

The most distinguishing feature of a solid, however, has to do with the orderly pattern of its atoms, which is repeated again and again. This orderly pattern, which we call **crystallinity**, characterizes most solids, including metals. Simple crystals are unique, however, because their shapes reveal to the naked eye secrets of their particular atomic structure.

Growing Crystals

For example, when water vapor freezes into solid snowflakes, the hexagonal shape of the ice crystals reveals the *ionic*—or open-packed—bonding pattern of oxygen and hydrogen atoms. If you heat water with potash alum, you can force the water to absorb more potash alum than it would normally at room temperature. This is called **supersaturation**. Crystals of the potash alum compound will form as the liquid cools, and their cube shapes reveal the close-packed arrangement of atoms in the alum.

How Crystals Grow

The orderly atomic arrangement of crystals is fascinating. But crystals do not suddenly spring into being; they *grow* into being.

Growing crystals do not draw nourishment from within, they grow from the outside—from materials presented to their surface. When growing an alum crystal in solution, the atoms of the compound diffuse through the water. When these atoms reach the crystal's surface, they join each other in an orderly pattern. New atoms extend this pattern outward.

Color of Crystals

Often identical atoms in a crystal are arranged without interruption. Sometimes different atoms that are the same size as the original atoms bump out the original atoms and replace them. So, impure atoms can invade the atomic structure of the crystal and create **mixed crystals** of dazzling hues.

Rubies, the precious stones, are mixed crystals of aluminum oxide which acquire their color by including a little chromium in place of aluminum.

Growing Crystals

You can grow crystals from supersaturated solutions. Some chemicals, when dissolved, produce large, colorless crystals. Other chemicals yield mixed stones of brilliant colors and fascinating shapes. This project requires planning ahead, since you will probably want to grow several groups of crystals and choose the best ones for your exhibit.

To learn about the shape and growth patterns of crystals, you'll want to produce large, clean-faced stones. This requires careful preparation of the supersaturated solution, and harvesting a single, unflawed *seed crystal* around which the larger crystal will grow. Be sure to have several clean jars with tight-fitting lids to prevent the solutions from evaporating. Begin tracing the openings of the jars on shirt cardboard. Cut out the circles and put them aside.

Clear Crystals

For your first **Rochelle salt crystal**, begin by pouring 100 cc of distilled water into a clean jar, and place the jar into a pot that's half full of water. Use this double-boiler arrangement on a hot plate. Warm the water in the jar to 50 degrees C (122 degrees F), and add 130 g of Rochelle salt, stirring until all the salt dissolves. Use the pot holder to remove the jar from the heat. Cover the jar and allow the liquid inside to cool. When it reaches room temperature, add a few more grains of Rochelle salt, shake the jar, and continue shaking it twice a day for the next two days. The added grains will not disappear. They will, in fact, grow slightly larger, but this procedure *primes* the solution for supersaturation later, and it prepares an ideal medium in which to grow a crystal.

After two days, carefully pour about an ounce (30 ml) of the solution into a small glass, and leave it undisturbed for another day. Watch for the tiny, individual seed crystals which begin to grow at the bottom. When a seed crystal reaches the size of a grain of rice, pluck it out with tweezers, and place it on a paper towel to dry. Harvest as many well-formed and unattached crystals as you can before discarding the solution in the jar.

To supersaturate the remaining solution in the jar, heat it to 50 degrees C (122 degrees F) again until the remaining grains of Rochelle salt disappear. Then, add another 9 g of salt, stirring until the liquid becomes clear. Remove the jar from the heat and allow it to cool.

Measure the jar from the top to about an inch (2.5 cm) above the bottom, and cut a piece of thread that length. Tie a seed crystal securely to one end of the thread, using a slipknot.

Take one of the cardboard circles you cut out earlier, and press it into the bottom of the jar lid. With a hammer and nail, punch three small holes in a small triangle in the middle of the lid, and loop thread through the holes so that the free end sticks out and allows you to adjust the string length inside the jar.

Screw on the jar lid, with the attached thread and seed crystal suspended inside the jar. A larger crystal will begin growing around the seed crystal in a few hours, and you will have a good-size specimen in three or four days. For a clean, symmetrical shape, keep the growing crystal from touching the bottom of the jar, adjusting the length of the thread through the lid.

Seed Crystal

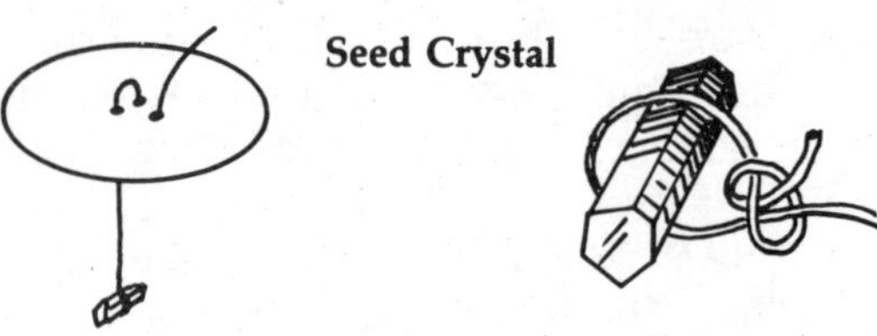

If you're satisfied with the shape and size of the crystal, remove it from the jar, and gently dry it with a paper towel. Remember, as clear and as brilliant as the Rochelle salt crystal appears, perspiration from your hands can damage it. Carefully examine the shape of the Rochelle salt crystal and make a sketch. Protect your finished crystal by wrapping it in a soft cloth or paper towel.

Purple Crystals

Potassium chromium sulfate (chrome alum) yields dark purple mixed crystals. Combining it with potassium aluminum sulfate (potash alum) allows you to vary the hue or create an unusual, layered stone.

The procedure for creating purple crystals is the same as that for clear crystals, but here you'll make two separate batches of solution—the potash alum first. And you'll need an extra jar handy for combining the liquids.

Pour 100 cc of distilled water into a clean jar, and place the jar in a pot half-filled with water. Warm the water in the jar to 50 degrees C (122 degrees F), and add 20 g of potash alum, stirring until the grains dissolve. Remove the jar, cover it, and allow the solution to cool, adding a few extra grains.

After a few days and periodic shaking of the solution, pour some of the liquid into a small glass and harvest a few seed crystals. To **supersaturate** the remaining solution, reheat the jar and add 5 g more of potash alum, stirring until the liquid becomes transparent. Remove the solution and pour half into a clean jar. Cover the jar to prevent evaporation.

Prepare the colored chrome alum solution the same way, adding 60 g of chrome alum to heated water. The solution will be dark blue green and difficult to see through, so continue to stir for a few minutes. Wait at least until the grains dissolve and the spoon glides smoothly over the jar's bottom. Remove the jar from the heat, cover it, and allow the solution to cool, adding a few extra grains. Supersaturate the solution by reheating it and adding 5 g more of chrome alum. Allow the liquid to cool.

Now carefully add the blue-green chrome alum to the half-filled jar of clear potash alum until you achieve the desired hue. Since this blue-green solution will produce a purple crystal, a darker solution will produce a more attractive stone than a lighter solution.

Suspend seed crystals from threads in all three jars—the clear potash alum, the dark chrome alum, and the combined solutions. After a few days, check your growing crystals (hold the dark chrome alum solution up to a light). After a week or when the crystals reach the size of a quarter, remove them from their solutions and allow them to dry. Examine and sketch their shapes.

You can transform the darkest crystals into a beautiful layered stone by suspending it in the clear solution. After about a week, a clear layer will form over the dark purple.

Red & Green Crystals

We will create two more crystals, one bright red and the other deep green, from solutions of potassium ferricyanide (red prussiate of potash)—and copper acetate monohydrate. Copper acetate monohydrate will be used with chemicals to produce our last crystal, brilliant blue.

For the **red crystal**, dissolve 93 g of potassium ferricyanide in 200 cc of warm water, cover the solution, and allow it to cool. Add a few more grains of the chemical, and allow the liquid to stand for a few days, shaking it at least twice a day. Pour a little of the solution into a glass and harvest the best seed crystals, tying a well-formed one to the thread. Supersaturate the remaining solution by reheating it while slowly adding 3 g more of the chemical. Then suspend the thread with its seed crystal inside the jar and wait for results.

For the **green crystal**, prepare a solution of 20 g of copper acetate monohydrate dissolved in 200 cc of water. If a film of undissolved chemical persists, add a few drops of glacial acetic acid, and stir well. cover this solution, and allow it to cool a few days—small seed crystals will develop spontaneously. Remove a well-formed seed crystal, tie it to a thread, and wait while a larger crystal with a more pronounced symmetry forms inside the jar.

Brilliant Blue Crystal

Creating a crystal of calcium copper acetate hexahydrate requires patience and careful preparation, but it's worth it. The stone is not only beautifully shaped, but its color will astonish you!

Calcium copper acetate hexahydrate is actually a compound made from calcium oxide and copper acetate monohydrate, used to create the green crystal.

Add 22.5 g of powdered calcium oxide to 200 cc of water, pour 48 g of glacial acetic acid into the mixture, and stir until the liquid becomes clear. If a small insoluble residue remains, filter the solution. For the filter, remove the bottom of a paper cup and replace the bottom with paper towels or a coffee filter. Dissolve separately 20 grams of copper acetate monohydrate in 150 cc of hot water. Mix the two solutions, cover the mixture, and allow it to cool for a day. If crystals do not appear spontaneously, let a drop of the solution evaporate and scrape off the resulting seed, tying it to the end of a thread as before. You should have a mature crystal in about 1 week.

If you're satisfied with your crystal collection, dispose of all chemical solutions by carefully placing the jars in a sink and rinsing them with cool water. Before rinsing the jars, you could pick out with tweezers the few remaining seed crystals at the bottom of each solution for future crystal growing. Wrap each seed crystal in a paper towel, and label it according to its chemical composition.

Classifying Crystal Shapes

Study the classification diagram of six basic crystal shapes. Now compare the sketches you made of your own specimens.

Group the finished crystals under their appropriate shape classification, providing descriptions and explanations. Display growing crystals suspended in solution alongside finished ones, dangling on their threads.

Crystal Shapes

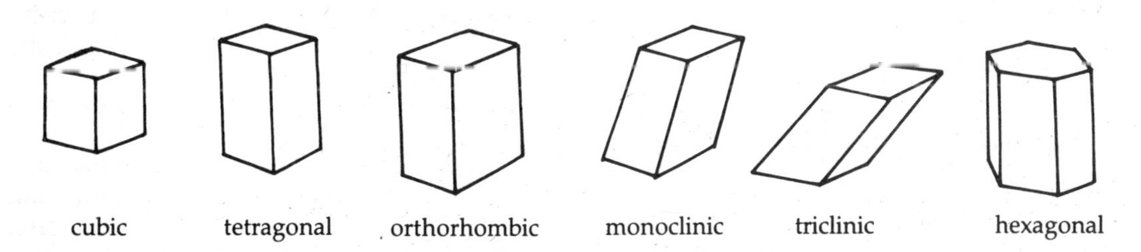

Solutions, Suspensions & Colloids

You Will Need

- 6 medium glass jars with covers
- 7 paper cups
- Eyedropper
- Measuring cup
- 1 cup (240 ml) of soil
- 1 cup (240 ml) of sugar
- 1 cup (240 ml) of tomato juice
- 1 cup (240 ml) of vegetable oil
- ½ cup (120 ml) of whole milk
- ½ cup (120 ml) of grape juice
- ½ cup (120 ml) of vinegar
- ½ cup (120 ml) of ammonia
- Measuring spoons
- ¼ teaspoon (1.25 ml) of table salt
- ¼ teaspoon (1.25 ml) of Epsom salt
- ¼ teaspoon (1.25 ml) of baking soda
- ¼ teaspoon (1.25 ml) of crushed antacid tablet
- ¼ teaspoon (1.25 ml) of cream of tartar
- Cone-shape coffee filter
- Blue food coloring
- Liquid soap
- Pen-light (flashlight)
- Black or dark poster board

Disappearing Act

Matter combines in infinite ways. A **solution,** like sugar dissolved in water, is one combination.

The way one substance dissolves in another is called its **solubility**. Since many variables are involved, study of solutions is complex. The dissolving substance, the **solute**, may be a solid, liquid, or gas; the medium into which it dissolves, the **solvent**, may be liquid or solid only. If it's difficult to imagine a solid mixing with another solid, think of the familiar mineral combination silver–gold ("white gold"). Chemists call such combined solids **solid solutions.** But since most solutions exist in liquid state, we'll concentrate on those.

Ordinary & Nonordinary Solutions

The study of solutions teaches chemists much about the nature of matter. Many chemical reactions take place *only* in solutions, and you can often tell what an unknown substance is by the solution it creates. For example, powdered sugar and baking soda look alike, but when you add equal amounts of each substance to equal volumes of water, the baking soda creates an *electrolytic solution*—a solution that conducts electricity. Powdered sugar, however, creates a nonconducting solution. Neither substance conducts electricity in the dry state; so, making solutions of each reveals important information about their chemical properties.

Based on the electrolytic potential of solutions, chemists classify them into two categories, ordinary and nonordinary. **Ordinary solutions** are chemically stable and neither acid nor base. They do not conduct electricity. Examples are sugar (sucrose) or ethyl alcohol dissolved in water. **Nonordinary solutions** are made from *salts* and have electrolytic properties.

Salts include a number of different substances, and they can be acidic, neutral, or basic when dissolved. Highly acidic salts, such as cream of tartar, make excellent conductors when dissolved, as do highly basic salts such as baking soda. Neutral salts, like table salt or Epsom salt, are also effective as conductors.

Acid Test

To determine the acid, base, or neutral quality of a nonordinary solution, we'll make a simple **indicator** from grape juice. Indicators are extremely useful to chemists.

First, add a few drops of grape juice to a paper cup containing ½ cup (120 ml) of vinegar. You'll see how the juice changes color in an acid. Add a few drops of juice to the ammonia to see the color change in a base. These colors function as your standards.

Place ¼ teaspoon (1.25 ml) each of table salt, baking soda, Epsom salt, crushed antacid tablets, and cream of tartar in individual paper cups and add ½ cup (120 ml) of water. Using an eyedropper, add several drops of grape juice to each cup, and observe how the color changes. As you noted at the beginning, grape juice turns *red in an acid* (vinegar) and *blue in a base* (ammonia). Each salt solution will turn a different color depending on its acidic or basic quality—red for cream of tartar, blue green for baking soda, green for antacid tablets, and pale purple (dilute grape juice) for the neutral solution of table salt and Epsom salt. You may want to further demonstrate the electrolytic properties of nonordinary solutions by rigging zinc–copper electrodes connected to a galvanometer.

Discovering the Solution

Solutions differ from other kinds of mixtures in that the solute breaks down into individual molecules within the solvent, resulting in a *homogenous* combination. A sample of sugar water taken from the top of a glass will have the same concentration of sugar molecules as an equal sample of sugar water taken from the bottom of the glass. Also, the only way to recover the solute (sugar) from the solvent (water) is to allow the solvent to evaporate, leaving the solute to concentrate into crystals.

In a jar, dissolve a cup of sugar in clear water and display it beside a jar filled with water alone. Can you see any difference between the two jars? Place a straw in the sugar water jar. Taste the liquid first from the top, then from the bottom. Does the sweetness vary?

Swirling Suspensions

Other forms of matter combine with liquids *without* dissolving into their individual molecules. This is called a **suspension**.

Add a cupful of soil to a jar of clean water. Cover the jar and shake it vigorously. Put the jar down and watch the soil particles swirl around in the water before they finally settle in the bottom of the jar. This may take a while, but eventually you'll have two layers—a layer of clear water at the top of the jar and a layer of soil at the bottom. We call this mixture a suspension because the particles of soil were *suspended* in the water temporarily before coming to rest at the bottom of the jar. Now, if you wanted to separate the water from the soil, you could carefully pour the water from the jar, leaving the soil undisturbed. This is called **decanting**.

If a suspension is made up of very small particles—like fruit or vegetable juices—it may take days or weeks for the particles to settle. In these situations decanting doesn't work, but filtering does. Place a coffee filter cone into the mouth of an empty jar, and pour a little tomato juice through it. What remains in the filter? What is the color and texture of the liquid beneath?

Colloidal Concoctions

The difference between solutions and suspensions has to do with homogeneity and the size of the particles. A solution consists of molecule-size particles, evenly distributed throughout. A suspension consists of particles large enough to be filtered or decanted from the liquid, particles that eventually concentrate in a layer. A third type of mixture, a **colloid**, resembles both solutions and suspensions in some ways but has properties of its own.

In a colloid, the particles are larger than molecules, but small enough to remain suspended permanently. Many household items are colloidal in nature. Paints, soaps, dyes, glues, jellies, coffee, and tea are all homogenous substances made of particles in permanent suspension. Mayonnaise and ice cream are also **immiscible-liquid colloids** with unusual features.

Although it's sometimes difficult to tell colloids and solutions apart, this test will help you. Pour ½ cup (120 ml) of whole milk into a jar of water. Mix it well. Place the jar against a dark background, such as black poster board, and shine a pen-light through the jar. You should see a beam of light travelling through the liquid. Shining the pen-light through the sugar-water and muddy water does *not* produce a beam. The colloidal mixture of milk and water consists of particles so tiny they act as little mirrors and reflect the light. This is called the **Tyndall effect,** after the scientist who discovered it. You can also see this effect when sunlight slants through a dusty room or headlights cut through fog.

Use your pen-light to test other liquids, and make a list of other familiar household colloids. If a liquid is too thick to shine a light through, dilute it with water until it becomes translucent.

Immiscible Liquids & Emulsifiers

The colloids mayonnaise and ice cream both consist of oil suspended in water. This is also true for Italian salad dressing, but the dressing's oil separates from its water. How do we account for the uniform texture of mayonnaise and ice cream? What do they contain that the salad dressing lacks? The answer is an **emulsifier.**

When two liquids do *not* mix together, we call them **immiscible.** Pour a cup (240 ml) of vegetable oil in a jar, and add water, dyed with blue food coloring. The oil forms beads which settle on top of the water. Some of the smaller beads combine to form larger beads until eventually you have a layer of oil. If you cover the jar and shake it, as you would shake salad dressing before using it, the oil breaks into tiny beads again. But if left undisturbed several hours, the beads of oil will combine, and the layer of oil will reappear.

Now add a few drops of liquid soap to the jar. Shake the jar as before and note the results. Instead of many small, but discernible oil beads, you now have a cloudy liquid. Even if left undisturbed, the oil has somehow been absorbed into the water and doesn't reappear. Soap acts as an emulsifier, surrounding the droplets of oil and preventing them from coming together.

Obviously, soap is an unacceptable emulsifier for foods. But other substances can serve the same function. In mayonnaise, egg yolk attaches itself to the oil droplets so that they do not combine and separate from the water-based vinegar. And as for ice cream, the problem is to keep the water droplets small enough to prevent grainy ice crystals from forming. Smooth ice cream has a secret emulsifying ingredient, gelatin.

Display your various jars with appropriate labels. You could mix up a fresh batch of nonordinary solutions and allow your viewers to identify each as acid, base, or neutral with the grape juice indicator.

Adsorption Chromatography

You Will Need

- Flexible plastic tubing, 80 × ¾ inches (203 × 1.9 cm) cut into ten 8-inch (20-cm) lengths
- Craft knife
- Stiff cardboard sheet 11 × 13 inches (28 × 32.5 cm)
- 2 stiff cardboard strips 2 × 10½ inches (5 × 26.2 cm)
- Hole punch
- 2 balsa-wood blocks for base, ¾ × ¾ × 3 inches (1.9 × 1.9 × 7.5 cm)
- Carpenter's glue
- Eyedropper
- 2 saucers
- Rubbing (isopropyl) alcohol
- Stick or wooden dowel, just large enough to fit inside tubes
- 1 of these substances: starch, (corn, rice, or white potato), powdered sugar, talc, calcium hydroxide (lime), calcium carbonate (powdered marble), diatomaceous earth
- Water-soluble ink—black, navy blue, peacock blue, violet, and sepia (brown)
- Food colorings: red, green, blue, and yellow
- Materials for plant chromatography: spinach leaves, grass, blackberries, strawberries, and colored flower petals
- Mortar and pestle
- Cotton
- Cotton swabs

Color Definition

Did you know that the green color of grass or spinach has blue, orange, and yellow in it? Or that black ink usually contains a red pigment? These are just two examples of many complex mixtures that can be separated into their components by a process called **chromatography**. The term comes from the Greek words *chroma*, meaning "color," and *graphy*, meaning "to write or describe." The process of chromatography "writes in color." As a liquid seeps downward through powdered material (adsorbent), some parts of the liquid cling harder to the medium than other parts. This results in a series of colored bands which tell chemists much about the nature of the liquid. Chromatography is also useful for separating valuable materials from natural or synthetic mixtures and for detecting impurities in foods and water.

Making a Stand

Assemble the stand for the adsorption tubes. Use the craft knife to make slots halfway through the two blocks of balsa wood. Insert the long edge of the 11 × 13-inch (28 × 32-cm) piece of cardboard into the slots so that it stands upright. Next, bend the two strips of 2 × 10½-inch (5 × 26.2-cm) cardboard lengthwise down the middle to form a 90-degree bend in each. Punch 10 holes, each ¾ inches (1.8 cm) in diameter, along half of each strip, and glue the solid side of the strip to the upright cardboard, separating them by about 5 inches (12.5 cm). Draw a number on the cardboard above each hole to help keep track of the tubes' contents.

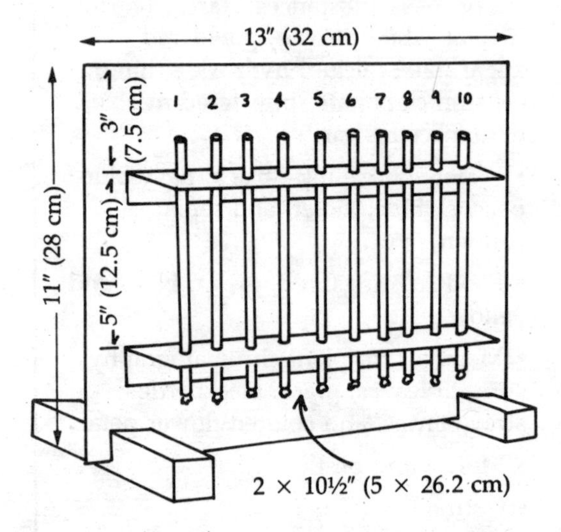

Test-Tube Holder

To keep the tubes from slipping through the bracket, wrap tape about 1 inch (2.5 cm) from the top of each tube. To keep the tubes from leaking powder, push a little cotton into the end of each tube.

Adsorbent Packing

Scoop a small amount of the selected adsorbent (such as cornstarch or powdered sugar) into the tubes, and carefully pack the adsorbent down with a wooden dowel. Continue scooping and packing until each tube is two-thirds full. Make sure each tube contains a uniform column of tightly compressed powder, necessary for the separation bands to stand out. If powder clings to walls of the upper section of the tube, clean it out with a cotton swab.

Chromatography Test

Put the tubes in the bracket and place the substance to be chromatographed in front of each tube. For the first five tubes, start with the inks—sepia first, then violet, peacock blue, navy blue, and black, in that order. For the next four tubes, use the red, green, blue, and yellow food coloring. For the remaining tube, try mixing together the food colorings in equal parts to create a new tint.

Fill two saucers with water, one for priming the powders before adding the solutions and the other for rinsing the eyedropper as you move from solution to solution.

Begin with the first tube. Add a few drops of water to wet the powder. Just when most of the water seeps into the powder, add 1 drop of sepia ink. Do not allow air into the column. Quickly rinse the eyedropper in the saucer of water. Now, just as the ink disappears into the column, add about 1 inch (2.5 cm) of clean water—or **developer**—to the tube. As water seeps into the powder, add more water so that no air can get into the flow column. Continue adding water until the bands of color stretch completely down the length of the tube.

Bands of Color

Repeat this procedure for the second tube using the violet ink, and so on, until you have chromatographs in all ten tubes. Allow the tubes to settle about 2 hours, noting results. The darker inks separate into many bands of surprising colors while the lighter inks separate into only a few. The red and blue food colorings do not separate, while the green food coloring separates into yellow and blue bands. The red, green, and blue mixture separates into three bands—blue, red, and yellow.

Primary Colors

Results may vary with the kind of powder used and the chemical composition of the inks and food colorings. In general, each solution separates into simpler or more **primary color components**. The primary colors—*red, blue, and yellow*—cannot be broken down, and they're the basis for all other colors in the spectrum.

Make a chart listing the color solutions, drawing their separation bands as accurately as possible with the colored pencils. Display the chart next to the samples.

Follow-up

For an interesting follow-up to the first group of chromatographs, repeat the entire procedure with the same solutions, but use a different adsorbent medium. Or use the same medium with a different developer. For instance, if you use calcium carbonate for the medium instead of cornstarch, the red, green, and blue mixture produces a blue zone at the top (instead of at the bottom), a green zone in the middle, and a yellow zone at the top. Passing the mixture through cornstarch as before, but using isopropyl alcohol for the developer, produces, from top to bottom—brown, red, yellow, and blue. Each of these chromatographs reveals different chemical details in the ink and food-coloring solutions. Reproduce the chromatographs with colored pencil and compare results on your chart.

Plant Chromatography

Plant chromatography reveals the component details of plant dyes. Liquify various plant parts—spinach leaves, blades of grass, blackberries, strawberries, and brightly colored flower petals—by adding a little water and grinding the mixture with a mortar and pestle.

Add these plant solutions to the tubes, again varying both the adsorbent medium and the developer. Record your results. The chemically simpler natural solutions should reveal *fewer* color bands. It should be obvious by now that there's more to the composition of substances—both synthetic and natural—than meets the eye.

Vegetable & Fruit Dyes

You Will Need

- Hot plate
- Small pot with cover
- Potato masher
- Strainer
- 2 large glass bowls
- Vegetables and fruits for making dyes—beets, carrots, squash, cabbage, onion skins, spinach, walnut shells, and berries
- 3-inch (7.5-cm) square samples of unbleached natural fibers—cotton, linen, silk, and wool yarn (tassels)

Natural Dyes

Before chemical dyes became inexpensive and popular in the mid-19th century, textiles were colored from natural sources. These unusual and delicate colors have not lost their appeal today, and sampling them will show you what hues people wore hundreds, if not thousands, of years ago. When you put your exhibit together, save a sample of each dye in a clear bottle, and display it next to several squares of dyed material.

Dying Directions

Place a few beets in the pot and add water. The amount is not critical, but generally the more beet or other vegetable, fruit, or shell you use, the darker the dye. Put the pot on a hot plate, cover it, and allow the water to boil about 20 minutes or until the beets are soft enough to be crushed with the potato masher. Carefully remove the pot with a pot holder, and pour the liquid through a strainer into a glass bowl. Discard the mashed vegetable matter in the strainer and pour the pure dye back into the pot, bringing it to a boil again. Turn down the heat so that the liquid does not evaporate.

Snip off a corner of the cotton samples so that you can tell them apart from the linen samples. Using the tongs, take one sample of each type of cloth and a wool tassel, and place them into the boiling mixture. Allow the samples to remain in the mixture for 20 minutes. Then remove the cloths and tassel with tongs. Rinse them in a bowl of clear water, and set them out to dry.

Repeat this dying procedure with each of the vegetables and fruits. You'll have to crush the walnut shells with a hammer before boiling them. As you remove your material and allow it to dry, record results in a chart like that shown. Unlike the beets, most of the other natural products create an almost colorless or transparent liquid.

In general, natural dyes produce pale colors on fabric. Also, the intensity of color depends on the kind of fabric used for a particular dye.

Gather other natural substances to turn into dyes. Try flower petals, lichens, tree bark, hickory chips, and natural clays.

Natural Material	Color of Natural Material	Color of Dye Produced	Color of Dyed Cotton	Color of Dyed Linen	Color of Dyed Silk	Color of Dyed Wool
walnut shells	tan	clear	brown			

Creepy Corsage

You Will Need

- Large jar with cover
- Ammonia
- Fresh flower in each of these colors—white, yellow, red, pink, and purple
- Construction paper

Undying Color

You'll enjoy the pale green color of these flowers, suspended upside down in a jar. This project demonstrates the presence of green **chlorophyll** under even the most vivid flower petal hues and the perishable quality of natural dyes.

Make a small corsage of flowers, a little narrower than the width of the jar, and tie them together with a short string. Attach another piece of string so that your bouquet hangs upside down in the center of the jar.

Remove the lid from the jar and punch a hole through the center with the nail. Push the end of the string through the hole and knot the end to prevent the string from slipping through. Pour just enough ammonia into the jar to cover the bottom (avoid inhaling the fumes) and screw on the lid. Your upside-down bouquet should sit, attractively suspended in midair.

Watch, as the flowers turn a ghostly green—the red, pink, and purple ones, that is. The white and yellow flowers remain the same. The ammonia destroys the pigments that give the reddish petals their bright colors, leaving only the faintest hint of green chlorophyll behind.

Dyes from natural products—plants, animals, and certain chemical compounds—are found in many paper and cloth products. Paper dyes consist of substances extracted from coal tar. The colors are brighter, but they're also relatively unstable and easily destroyed.

Paper Bouquet

Replace the flowers with strips of construction paper in various colors, bunched together and suspended from a string. Dip your paper bouquet in water before inserting it into the ammonia jar. Notice again how the more perishable reds and blues fade while the yellows and oranges remain.

Shadow Pictures

Make shadow pictures using construction paper, unusually shaped objects, and sunlight. Place the objects on the construction paper, and leave the paper on a windowsill undisturbed for several days. The paper surrounding the objects will gradually fade to white or gray, leaving "shadows" in color.

pH of Cosmetics & Cleaners

You Will Need

- Head of red cabbage
- Electric blender
- Small bowl
- Plastic cups
- Plastic spoons or stirrers
- Large wire strainer
- Eyedropper
- 3 different water-soluble samples each from 4 household groups—food, cleaning products, medicines, and cosmetics and personal hygiene products.

Everyday Chemistry

Like chemists, we naturally categorize everyday substances by their functions. Food, medicines, cleaning agents, cosmetic and personal hygiene products all have an appropriate place and function in our lives. Professional chemists, however, determine the chemical composition of a product and can explain why a substance might only be appropriate on the skin, or why a sweet-smelling bath product should not be swallowed.

Acid or Base

Chemists use the pH scale to identify whether a substance is acid, neutral, or base. **Acidic** substances have a pH from 1 to 6, **neutral** substances have a pH of 7, and **basic** substances have a pH of 8 to 14. In this project we'll test the pH value of various household products.

First, we'll mix the **indicator** used to test samples; it will change color according to the sample's particular pH value. Remove 4 cabbage leaves and put them into a blender with 1 cup (240 ml) of water. Blend until the mixture becomes deep purple. Pour the liquid through a strainer and into a small bowl. Place it aside.

Create a data table to display beside your results. This table will help you draw important conclusions about your experiment.

pH Test

Choose three different samples from each substance group, placing each sample in a plastic cup. Dissolve the sample in a little water if it is not already in liquid form.

Fill the eyedropper with the red cabbage indicator, and carefully add 2 or 3 squirts of fluid into each of the plastic cups. Note any color change, and write the substance's name in the proper column on the data table. On a separate chart, group the products you tested according to their functions, indicating the color and pH range of each.

Most *foods* are generally acidic (lavender or pink), *cosmetics* tend to be neutral or slightly acidic (blue or lavender), and *cleaning products* are either acidic or basic depending on their functions. The pH for *medicines* varies—acid for aspirin, base for antacids.

Arrange selected items, such as soap, alcohol, bleach, and fruit and vegetable juices, with empty cups on your display table. Invite viewers to participate in testing. See if you can predict the results.

Testing pH Values

Acids		Neutral	Bases	
Pink (pH 2–3)	Lavender (pH 4–5)	Blue/Purple (pH 7)	Yellow (pH 8–9)	Green (pH 10–11)

Detecting Vitamin C in Fruits & Vegetables

You Will Need

- 2 cups (480 ml) distilled water
- Hot plate
- Cooking pot
- 2 medium jars with lids
- Small bowl
- 100-ml beaker
- 500-ml beaker
- Eyedropper
- 100 mg vitamin C tablet
- Tincture of iodine
- 2 cups of each juice—orange, pineapple, grape, tomato, and green pepper

Vitamin C in Our Diet

One of the most important vitamins in our diet is **ascorbic acid,** better known as vitamin C. Found in many fruits and vegetables, it's particularly abundant in citrus fruits. Vitamin C is also easily damaged or destroyed. Unlike other vitamins, it cannot be stored long in our bodies and must be replenished each day.

Cornstarch Testing Solution

This project compares vitamin C concentrations in various fruit and vegetable juices. We'll also test how heat affects the vitamin.

Refrigerate a glass of distilled water until it's quite cool. Add 1 teaspoon of cornstarch to the glass, then pour the water into a pot and boil it. Carefully remove the pot with a

pot holder, and allow the water to cool. Fill the eyedropper with the cooled water, and add 20 drops of it to a jar of fresh distilled water. Cover the jar and refrigerate.

Making a Vitamin C Standard

Place a 100 mg vitamin tablet in a small plastic bag and crush it with a hammer into a powder. Fill the 100-ml beaker with 50 ml of distilled water. Add the powdered vitamin C, mixing well. Then fill the beaker with water to the 100-ml mark.

Pour this solution into a jar and screw on the lid. Wrap masking tape around the jar's surface and refrigerate it. Rinse out the 100-ml beaker.

Titrating the Vitamin C Standard

To determine the concentration of a substance in a solution, chemists use the technique called **titration**. This means *adding*, in precisely measured amounts, a reacting agent of known concentration until the solution changes color, indicating a chemical reaction.

To titrate your vitamin C standard, pour

200 ml of distilled water into a 500-ml beaker. Add 10 drops of iodine and mix well. Place 25 ml of this solution in the clean 100-ml beaker. Now, fill the eyedropper twice with the refrigerated cornstarch solution, and add it to the iodine solution. The combined iodine and starch molecules turn this mixture blue. Rinse the eyedropper in the bowl of water to remove traces of the cornstarch solution.

Remove the jar of **vitamin C standard** from the refrigerator. Add it, drop by drop, to the blue solution until the color changes to clear. Adding vitamin C to the mixture breaks down the iodine–starch combination and the blue color disappears. Record the number of drops it takes to do this; then, rinse the beaker and eyedropper.

Titrating Juices

Add 25 ml of iodine solution to the 100-ml beaker as before. Add two squirts of iodine to create the blue mixture again. This time, add orange juice, drop by drop, until the solution turns clear. Record your results.

Repeat the procedure with grape, pineapple, and tomato juice. Then, pulverize a green pepper by placing it in a blender, and use some of that juice, too. Each time, record the number of drops it takes to turn the solution clear.

Since adding vitamin C to the blue solution breaks down the iodine–starch combination, the *more drops* of a substance containing vitamin C you need to turn the solution clear, the *weaker the concentration* of vitamin C in that substance. Of all the juices, which seemed to contain the highest concentration of vitamin C? Results may surprise you: green pepper contains more vitamin C than orange, pineapple, grape, or tomato.

Vitamin C & Heat

Now heat 1 cup each of orange, pineapple, grape, green pepper, and tomato juice to boiling, and repeat the titration procedure. No amount of juice affects the blue iodine–starch solution. Heat has completely destroyed the vitamin C in all five juices.

Bacterial Content of Milk

You Will Need

- 2 test tubes with rubber stoppers
- 2-hole test tube stand
- 2 medium glass jars
- Saucepan
- Hot plate
- Tongs
- Thermometer
- Calibrated (cc) medicine dropper
- Methylene blue solution
- Refrigerated milk
- Standing milk
- 3 large poster boards

Methylene Blue Test

Measuring the bacterial content of milk with the **methylene blue test** is just one procedure scientists use to monitor the quality of our foods. The blue dye indicates the presence of *dissolved oxygen* in the milk. Since growing bacteria also require this oxygen, the time it takes the blue to disappear conversely indicates the amount of bacteria present.

Your Experiment

You'll need precisely measured amounts for accurate results; so use a calibrated eyedropper and calibrated test tubes. Most pharmacists stock methylene blue.

Sterilize the test tubes. Bring water to a boil in the saucepan. Grasp the test tubes with tongs and carefully drop them into the water along with their stoppers, removed from the tubes. Avoid splashing. Allow the tubes to boil 1 minute. Then remove them with tongs, and place them on a pad of paper towels to cool and dry.

Place the test tubes in the stand. Using the calibrated eyedropper, place 9 cc of refrigerated milk into both tubes. The first tube functions as our test sample, and the second tube is our control, just for comparing color. Add 1 cc of methylene blue to the first test tube, noting the time. Place the stopper on the test tube, and shake the tube until the blue mixes thoroughly with the milk. Place the stopper on the second tube.

For accurate results, keep both samples at 98.6 degrees F (37 degrees C). Place water in the saucepan, and slowly heat it on the hot plate. Fill both glass jars three-quarters full with water, and place them in the saucepan. Add water to the saucepan to the water level of the jars. Place the thermometer water the jars, and slowly heat the water until you reach 98.6 degrees F (37 degrees C). Make any necessary adjustments in the heat of the hot plate to maintain this temperature.

Place a test tube in each jar and allow them to remain awhile. Check the tubes every half hour for the first 2 hours, and once an hour after that. If the blue color disappears in streaks, gently stir the sample with the thermometer, but first sterilize it with a little alcohol and wipe it clean with sterile gauze.

When the milk containing the methylene blue becomes as white as the undyed milk, the test is over. Note the time it took the color to disappear.

Results

Consult the Bacterial Count chart to determine the quality of the milk you tested. If it took 5½ hours for the color to disappear, the milk from which you obtained samples is of excellent quality and quite safe to drink.

For quicker and more dramatic results, try repeating this demonstration substituting the standing milk for the refrigerated milk. Milk left at room temperature overnight will take just a half hour for the color to disappear.

Display milk samples in their test tube holders along with your data and copies of your charts. Milk quality concerns everyone.

See the charts Bacterial Count and Graded Milk Standards to evaluate acceptable and unacceptable levels of contamination. The Graded Milk Standards chart represents the graded milk standards set by the U.S. Public Health Service. Other countries may have other standards.

Bacterial Count

Time for Milk to Regain White Color	Quality of Milk	Approximate Number of Organisms per cc of Milk
over 8 hours	excellent	variable, but very good
5½ to 8 hours	good	under ½ million
2 to 5½ hours	fair	½ to 4 million
20 minutes to 2 hours	poor	4 to 20 million
under 20 minutes	very poor	over 20 million

Graded Milk Standards

Type of Milk	Grade of Milk	Approximate Number of Organisms Allowed per cc of Milk
raw	certified	under 10,000
raw	A	under 50,000
raw	B	under 200,000
raw	C	under 1,000,000
raw	D	over 1,000,000 safe for cooking only
pasteurized	A	under 30,000
pasteurized	B	under 50,000
pasteurized	C	over 50,000 safe for cooking only

Decomposition of Natural Plastic

You Will Need

- 2 glass jars
- 4 ounces (120 ml) of whole milk
- 1 teaspoon of vinegar
- Small plastic charm
- 2 clay flowerpots
- Soil

Making Plastic Naturally

Using What Was Natural

Chemists have learned to create artificially—or **synthesize**—many materials once found only in nature. Although plastic might seem like an entirely synthesized material, natural plastics made from plants and animal fat were used over 100 years ago. A 19th century lady probably had a small plastic button, pin, or hair comb, and an 18th century farmer's diary provides a recipe for "hard paint for barns," made from milk mixed with cow's blood.

Decomposition

The difference between natural plastics and today's petroleum products has to do with **decomposition**. Plastic made from plants and animals breaks down eventually, leaving no toxic or polluting materials behind. **Petroleum products**, however, do not decompose, and they pose a threat to the environment. Recycling petroleum-based plastics has, in recent years, addressed this concern.

Homemade Plastic

You can create your own homemade plastic with everyday materials. The look and feel of it will surprise you!

Pour 4 fluid ounces (120 ml) of whole milk into a saucepan and heat. The milk will first boil, then separate into tiny particles (curds) and a clear liquid. Using the pot holder, slowly pour off the liquid from the pot into one of the glass jars. Then spoon the curds into the other glass jar.

Add 1 teaspoon (5 ml) of vinegar to the curds, and allow the mixture to stand about two hours. The individual curds will seem to "melt" into a solid, yellowish mass at the bottom of clear liquid. The mass consists of fat, minerals, and the protein **casein**—a string-like molecule that bends like rubber until it hardens. Pour out the liquid, and carefully remove the mass from the jar, kneading until it has a doughy consistency.

Mold your plastic into any shape or shapes, placing them on a sheet of waxed paper to dry overnight. Examine your finished pieces carefully. How does natural plastic compare to the petroleum product?

Exhibit Possibilities

For your exhibit, detail each stage of plastic-making with photographs and explanations, displaying the bottle of curds before and after adding the vinegar.

In a second part of your exhibit, compare the differences in decomposition between petroleum and natural plastics. Bury a sample of each in a flowerpot filled with wet soil. After a week, remove each sample and compare them, noting any changes in color, pliability and texture. You will find that the natural plastic shows evidence of decomposition while the petroleum product remains unchanged. Mount the samples with your conclusions and suggestions.

The Future of Plastic

In recent years, chemists all over the world have developed various types of **biodegradable plastic** by adding substances that can be attacked by light, bacteria, or other chemicals. Adding starch to plastic helps it break down because of starch-eating bacteria in soil. **Chemically degradable** plastics, useful as protective coatings for metals, can be dissolved with certain solutions. Surgeons also use chemically degradable plastics designed to dissolve in body fluids. **Photodegradable** plastics contain substances that slowly disintegrate when exposed to light. Farmers use such plastics to temporarily cover crops and retain heat.

Recently, the Japanese have developed a low-cost biodegradable plastic made from the shells of shrimp. They extract a substance from the shells called **chitin,** also found in human fingernails, and combine it with silicon. The material produced, **chitisand,** is stronger than petroleum-based plastics. Not only does it break down in soil, but it has fertilizing properties as well!

Earth, Clouds & Beyond

The Biggest Magnet

You Will Need

- Sensitive pocket compass
- Notebook

Soup Can & Earth—Poles Apart

You can easily discover the enormous effect of the earth's **magnetic field**.

Place 2 soup cans, which have been sitting on the shelf a few days, 6 inches (15 cm) apart on a flat surface. Turn the second can upside down. Hold a compass beside the top of the first can, and observe how the needle deflects, attracted towards the can. Now slowly move the compass down along the can's side, and observe the needle swinging away until it points in the opposite direction by the time you reach bottom. You've discovered a *miniature magnetic field* around the can—a replication of the earth's magnetic field. Like the earth, the can has a north and south pole, and is actually a *weak magnet*.

Upside-Down Results

Repeat the procedure for the second, upside-down can. Notice how the compass points south near the top and north near the bottom, in a mirror image of the first can. If you leave this can undisturbed a few days, it will be remagnetized and the poles will reverse.

Magnetized Metals

Everything *metallic* on earth has been magnetized to some degree. You can test this by holding your compass to the tops and bottoms of assorted metal objects and gauging the needle's deflection. Even very large objects, like refrigerators, safeboxes, lampposts, cars, and trailers, have magnetic fields. Also, some materials found in the earth are *naturally magnetic*, like iron ore.

List objects you examine, type of metal (if you can identify it), and degree of needle deflection. Note when the compass begins to swing in the opposite direction. Is it always at the mid-point of the object? Do you find any difference in needle deflection between stationary objects, like metal fence posts, and moving ones, like bicycles? Display results on a chart, compare data, and draw conclusions.

Seismograph

You Will Need

- Base for large outdoor umbrella, either concrete or sand-filled
- 4 cinder blocks 16 × 8 × 6 inches (40 × 20 × 15 cm)
- Plaster of Paris
- Inexpensive clock, windup or electric
- Round tin can 3½ × 3 inches
- Round tin can 5⅕ × 3 inches
- ¼ × 2-inch (.62 × 5-cm) angle brace
- ⅛-inch (.31-cm) square steel rod 3 inches (7.5 cm) long
- Strip of 30-gauge aluminum 41 inches (102.5 cm) long
- 2 brass plates 1 × 3 × ⅛ inch (2.5 × 7.5 × .31 cm)
- Brass plate ¾ × 2½ × ⅛ inch (1.8 × 6.25 × .31 cm)
- Brass plate ¾ × 5 × ⅛ inch (1.8 × 12.5 × .31 cm)
- Brass strip ½ × 10 inches (1.25 × 25 cm)
- Brass strip ½ × 4½ inch
- Brass rod 2 × 3/16 inch (5 × .47 cm)
- Brass rod 2¼ × ⅜ inch
- Brass rod 4 × ⅜ inch
- Brass tubing 6½ × ⅜ inch (16.25 × .93 cm)
- ¾-inch (1.8-cm) plywood—1 piece 8 × 22 inches (40 × 55 cm), 1 piece 4 × 5½ inches (10 × 13.75 cm), 2 pieces 2½ × 5 inches (6.25 × 12.5 cm), 2 pieces 3 inches (7.5 cm) square
- 2 × 4 wood 3 inches (7.5 cm) long
- Thin iron wire, 14- to 16-gauge
- Table vise
- Hacksaw
- Pliers with wire clippers
- Soldering iron
- Tin snips
- Drill
- Tap (for threading holes)
- Assorted nails, including 2 finishing nails
- Assorted nuts, including lock nut
- 2 machine bolts 1 × ¼ inch
- Square washer
- Flexible shaft universal joint (coupling for radio condenser)
- Steel phonograph needle
- 2 sewing needles
- 3 fabric rivets (metal snap buttons)
- Wooden matchstick
- Carpenter's glue
- Aluminum foil

Earth Rumblings

No matter where you stand on earth, the ground trembles at this very instant beneath your feet. Our planet rumbles with geological activity—mountains push upwards, canyons sink, continents ride the backs of fire-fringed tectonic plates. By constructing a relatively uncomplicated **seismograph,** you can record some of this activity.

This homemade earthquake detector can record shock waves at a distance up to 1,000 miles (1,600 km). Yet your seismograph will remain unaffected by traffic or construction vibrations.

Earthquake Basics

Seismologists classify waves into two types—**compression** (P *waves*) and **shear** (S *waves*). Compression waves create an up-and-down vibration through the layers of the earth, while shear waves cause the layers to move from side to side. Shear waves cause far more destruction than compression waves and register clearly on seismographs. That's why your instrument will be able to record distant shear waves and yet remain unaffected by any compression waves of cars, trucks, and heavy machinery.

Your Seismograph

The model here has three main components—adjustable post, weighted pendulum containing the stylus, and revolving drum. The instrument works by scratching a straight line onto a piece of smoked aluminum foil fastened to the drum. When the ground trembles, the pendulum swings and the stylus scratches a *wavy* line on the foil. The size of the waves reflects the intensity of the tremor.

Post & Pendulum

Drill two ¼-inch (.62-cm) holes at opposite ends of the angle iron post, following the diagram. The top of the post is the side where the hole comes closest to the edge. Place the bottom of the post in the pipe of the umbrella base. Make sure the post doesn't lean. If it fits snugly, you can leave it as is. But if the post moves around, mix a small amount of plaster and pour it into the pipe to anchor the post in place.

For the extension arm or pendulum, take the 41-inch (102.5-cm) strip of aluminum and form it into a tapered channel by folding up the sides as indicated. Make the narrow end into a Y shape by cutting a 2½-inch (6.25-cm) notch in the strip and spreading the sides. With a small nail, punch two holes at the ends of the Y to contain the stylus.

Again, looking at the post, notice on the diagram the three adjustment points necessary for fine-tuning the seismograph (A, B, and C). The upper, or wire, adjustment (A) consists of a threaded brass rod bent into an L shape and pierced with a small hole. Screw it into a brass plate bent back at a little more than 45 degrees. A lock nut behind the bracket keeps the adjustment rod from changing position.

Adjustment B, for shifting the upper support sideways, consists of a U-shape brass plate, with a threaded brass rod running through the ends. Leave the U unthreaded. It serves only as a bearing for the rod, which is kept from slipping out by two nuts soldered to the rod after being tightened against the U.

Attach both the U- and the L-shape brackets to the post with ¼ × 1-inch machine bolts.

Finally, adjustment (C) is made of the ⅜-inch (.93-cm) brass rod threaded full length, with a hacksaw-cut slot in the rear for a screwdriver. At the other end of the rod, solder the female side of a fabric rivet, flat side against the rod. The phonograph needle of the stylus fits into the rivet's hollow part, allowing the stylus to pivot freely. Place nuts before and behind the rod, after inserting it in the lower hole of the post. Tighten the nuts so that the fabric rivet protrudes halfway from the surface of the post.

The Weight

Construct the weight, or **bob**, from a tin can 3½ inches (8.75 cm) high and 3 inches (7.5 cm) in diameter. Place the 3-inch (7.5-cm) piece of 2 × 4 wood inside the bob, and hold

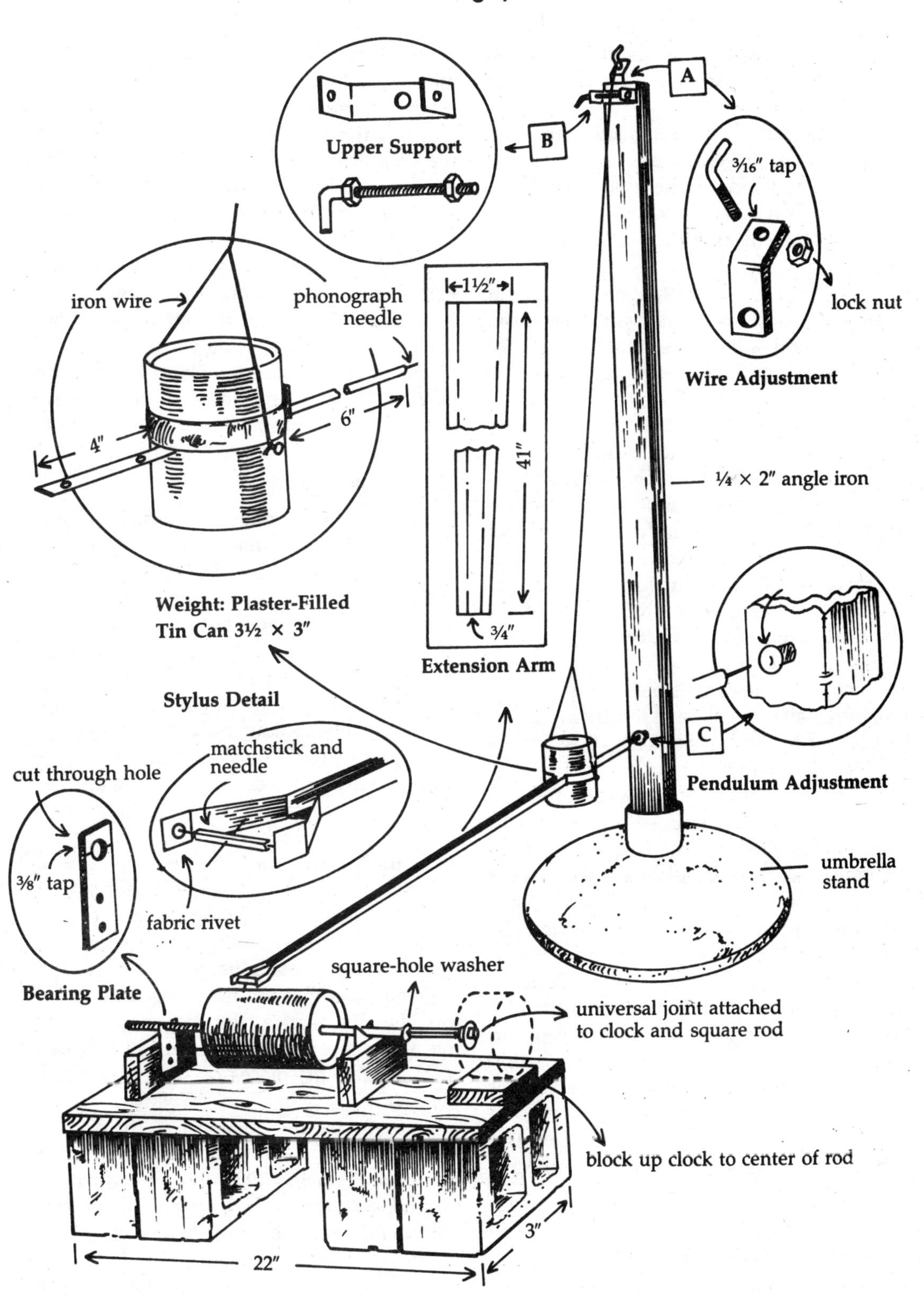

Seismograph Assembly

it there to support the walls of the tin can and to prevent it from rolling during the following operation.

In the middle of the can's side, drill and tap a ¼-inch (.62-cm) hole. Then, in the middle of each side of the bob, at right angles to the rod hole, drive a small finishing nail so that the head protrudes about ¼ inch (.62 cm).

Thread a 6 × ¼-inch brass rod at one end to screw into the bob. Using the hacksaw, cut off the end of the rod so that the distance to the middle of the bob is 5¾ inches (14.37 cm).

Secure the rod in the table vise, and carefully drill a hole in the end, 9/16 inch (.93 cm) deep and slightly wider than a steel phonograph needle. Apply a little solder around the needle to secure it in the hole. The rod should now measure 6 inches (15 cm) from the tip of the needle to the center of the bob.

Screw the ¼-inch rod into the side of the bob until it reaches the center. Prop up the end of the rod on a brick so that it sits perpendicular to the can's side. Straighten the finishing nails as well, making them as perpendicular to the can's sides as possible.

Mix just enough plaster to fill the can, pouring it slowly over the rod and nails to avoid disturbing their position too much. After you've filled the can, correct any shifting in the rod or nails. Allow a few hours for the plaster to dry before moving the can.

For attaching the pendulum to the bob, begin by drilling 2 holes in a strip of brass, ½ × 4½ inches (1.25 × 11.25 cm). Bend ½ inch (1.25 cm) at the end into a right angle. Solder this end to the bob, directly opposite the ¼-inch (.62-cm) rod hole. Fasten this piece to the pendulum with 2 small bolts.

Now make a sheet brass clamp ½ × 10 inches (1.25 × 25 cm). Wrap it around the bob and over the soldered strip, securing it with a small screw.

Constructing the Stylus

Too much friction will prevent you from obtaining accurate data. To minimize it, you'll use a counterbalanced needle as the centerpiece of your stylus. Cut down a wooden matchstick to ¾ inch (1.8 cm). Cut two small sewing needles in half with wire clippers, and put the pointed halves in the ends of the stick. Run another small needle through the center of the matchstick to form a cross. The point of this needle traces the line on the drum.

For pivot bearings, solder fabric rivets, with female sides facing each other, to the ends of the Y. Mount the stylus pivot between them. After you've finished the drum, adjust the counterbalanced needle by moving it back and forth until it barely touches the aluminum foil.

Drum & Base

Make the drum from a 5½ × 3-inch tin can. Use a discarded condensed milk or juice can because you need both the top and bottom of the can intact. Drill a hole through the center of the top and bottom, and mount the can on a 16½-inch (41.25-cm) axle made from ⅜-inch (.93-cm) brass tubing. Use the tap to thread 5½ inches (13.75 cm) on one end of the tube. Solder a square-hole washer (the square should be slightly more than ⅛ inch [3 mm] on a side) at the other end.

Make bearings for the axle from two strips of 1 × 3-inch (2.5 × 7.5-cm) brass. Near the narrow end of one strip, drill a 7/16-inch (1-cm) diameter hole, and thread it to fit the axle threads. In the other strip, drill a ⅜-inch (.93-cm) hole, but leave it unthreaded for a plain bearing. Cut off the bar's ends so that the holes are cut in half, too. Finally, screw the bearings to the mounting trestles made from 2½ × 5-inch (6.25 × 12.5-cm) pieces of ¾-inch (1.8-cm) plywood.

Glue the trestles edgewise to the 8 × 22-inch (20 × 55-cm) plywood base, separating them by 11 inches (27.5 cm). Or turn the base upside down and attach the trestles with nails. Straddle the drum across the trestles and rotate it. Notice how the drum moves sideways in the direction of the threads on the axle. This movement guarantees that the stylus will trace a continuous spiralling line. Driven one revolution per hour by the electric clock's minute-hand gear, the drum rotates slowly enough so that the foil will last awhile without being changed.

To connect the clock to the drum, you need a ⅛-inch square steel rod to fit into the square washer soldered to the unthreaded end of the axle. If you use an electric clock, make sure you unplug it. Remove the hands and cut a hole in the clockface just large enough to expose the driving mechanism. Block up the clock with 2 pieces of 3-inch (7.5-cm) square plywood until the driving shaft sits level with the axle.

Attach one side of the flexible coupler (universal joint) to the minute-hand shaft, and the other side to the end of the square steel rod. Solder the minute hand onto the rod, close enough to the clockface so that it clearly indicates the hour. Make sure the drum, driven by the clock, rotates away from the stylus needle.

Finally, arrange the four cinder blocks to function as a foundation for the plywood base. The drum should sit at about the same level as the base of the post.

Hanging the Pendulum

Over the finishing nails at the sides of the bob, attach a looped stirrup of thin iron wire (a guitar E string works well). Tie the suspension wire to the middle of the stirrup, bringing the point of suspension to the middle of the bob.

Thread the end of the suspension wire through the hole in adjustment rod A, and rest it between threads on adjustment B. Turn A until the pendulum, with the needle tip in the bearing of C, is horizontal from the bearing to the bob center. The bob should come to rest with its rod at a right angle to the face of the post. Make any necessary adjustments by turning the rod of B.

Preparing the Foil

Your seismograph records tremors with a stylus needle scratching waves on the surface of blackened aluminum foil. Cut a strip of foil slightly wider than the drum and long enough to wrap completely around it. Bend the foil over the drum's sides, and attach it with tape. Be extra cautious to avoid wrinkling the foil where it covers the drum.

Perform this next step outdoors with adult aid. Hold the drum by the axle and slowly rotate it over a candle flame until you cover the foil with a layer of soot. Holding the drum over the tip of the flame will provide best results.

Replace the drum on its bearings, and position the platform so that the stylus needle sits on the drum edge farthest from the clock. Wind or plug in the clock to start the drum revolving.

Cutting a Record

The seismograph makes an excellent science project because you can simply let it operate and check results periodically. Remove the foil when the needle reaches the end of the drum and replace it with a fresh piece. To preserve your records for display, coat them with a thin layer of spray varnish.

Sling Hygrometer

You Will Need

- 2 identical thermometers with rigid backings
- ½ × 6-inch (1.25 × 15-cm) dowel, cut down as needed
- 1½-inch wood screw
- Washer
- Carpenter's glue
- Cotton
- Drill

Weather Basics

Weather, good or bad, affects nearly every aspect of our daily lives. The science of weather prediction—**meteorology**—probably began when someone first lifted a finger to determine the direction of the wind. Weather-measuring devices are still relatively uncomplicated. You'll need about 30 days to collect data.

What's Humid?

Humidity—the amount of water vapor in the air—makes all the difference in determining whether a warm day is pleasant or muggy and a cool day refreshing or clammy. Meteorologists find the figure for **relative humidity** by comparing the amount of moisture in the air with the amount it could hold if it were *saturated* at the same temperature.

Reading Double

This sling hygrometer will give two simultaneous temperature readings. Using the chart will help you determine relative humidity with reasonable accuracy.

Attach two thermometers back-to-back with carpenter's glue. If the thermometers have plastic backings, use epoxy glue instead. Allow the glue to dry.

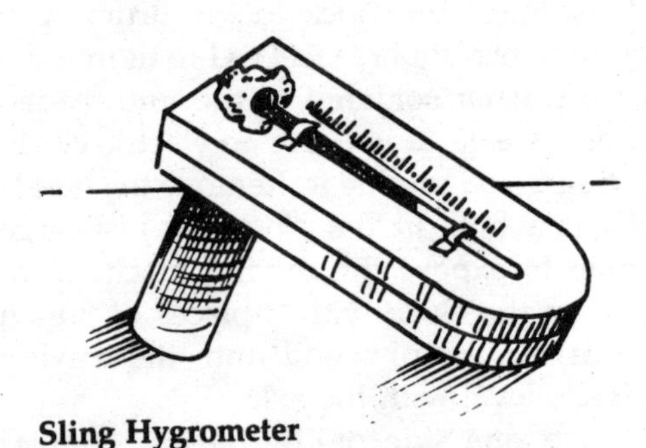

Sling Hygrometer

Drill a hole through the two thermometers at the top of the mountings. Then drill a hole at the wooden dowel's top. Attach the thermometers to the dowel with a 1½-inch (3.75-cm) wood screw, placing the washer between screw and dowel. Don't tighten the screw all the way; the thermometers should swing freely around.

Cotton Test

Stuff cotton under the top thermometer's bulb, fastening it with two thumbtacks if necessary. Moisten the cotton with water. Holding the dowel, whirl the hygrometer around for 30 seconds before recording the temperature on both thermometers. The wet bulb should now read a *lower* temperature than the dry bulb, since heat is required to evaporate the water.

Using the chart, locate the dry-bulb temperature at the bottom. Locate the wet-bulb temperature at the top, where indicated by the diagonal lines. Trace these two lines until they intersect. To find the percentage of humidity, follow the nearest curved dotted lines to the right of the chart. If the dry-bulb temperature were 70 degrees F (21 degrees C), for example, and the wet-bulb temperature 62 degrees F (17 degrees C), you'd find the relative humidity would be 60 percent, well within the comfort zone.

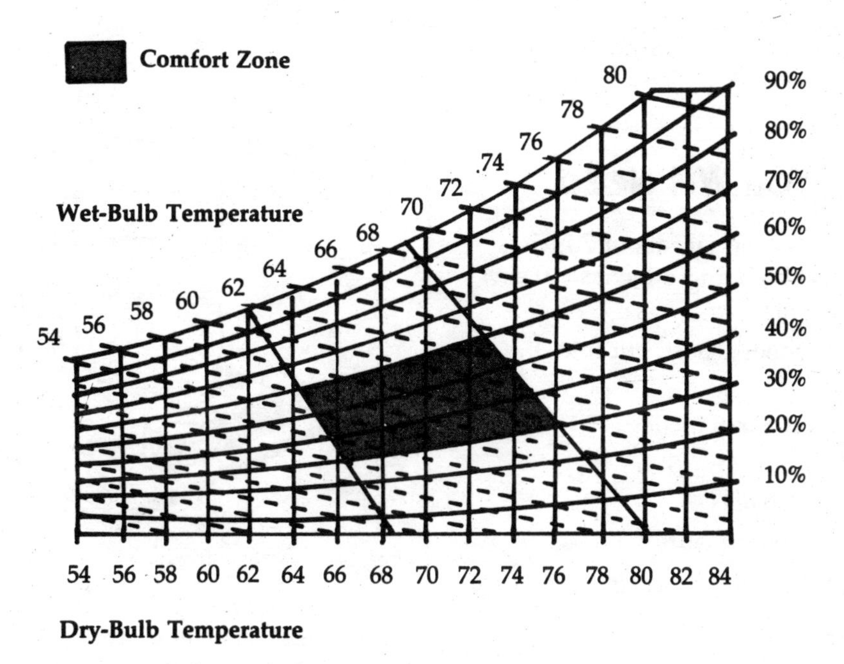

Determining Humidity

Rain Gauge

You Will Need

- Straight-sided jar, about 7 inches (17.5 cm) high and 4 inches (10 cm) in diameter
- 1 × 1 wooden post, cut into 3 sections
- Plywood 6 × 6 inches (15 × 15 cm)
- Medium nails
- Cloth tape measure
- Clear cellophane packing tape
- Electrical tape

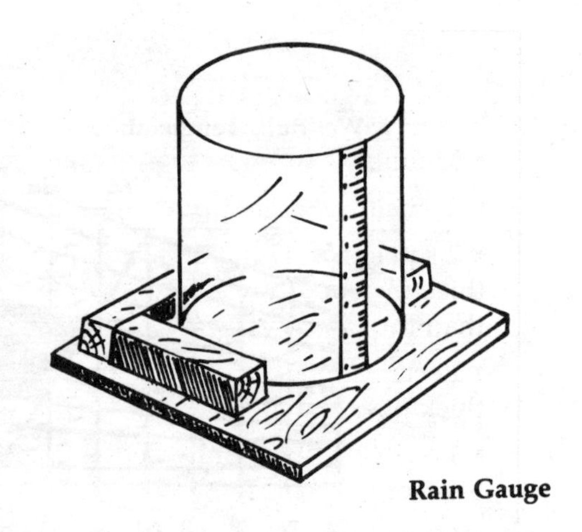

Rain Gauge

Precipitation

Precipitation, the amount of rain, snow, sleet, or hail that falls within a 30-day period, is an important part of the weather picture. A simple rain gauge measures the amount of precipitation for any season of the year.

Constructing a Rain Gauge

Place the tape measure against the jar and measure it from bottom to top. If the jar is 7 inches (17.5 cm) high, cut the tape measure at that length, and attach it, with wide cellophane packing tape, to the side of the jar. Make sure the tape measure lies flat against the jar's surface.

Position the jar in the center of the 6 × 6-inch (15 × 15-cm) plywood base, and carefully draw around the base with a pencil. Remove the jar, and use a ruler to measure the circle's diameter. Draw parallel lines the same length as the diameter on opposite sides of the circle. Join the parallel lines with a perpendicular line across the top so that you have a circle enclosed by three sides of a square.

Cut two pieces from the 1 × 1 post the same length as the parallel lines. Cut a third piece 2 inches (5 cm) longer. Assemble the pieces as shown in the diagram, using nails to attach them to the plywood base.

The wooden pieces should form a cradle you can slide the jar into to keep it from tipping over. Make sure the jar fits snugly, with the tape measure side facing you. If the jar twists easily from side to side, put a piece of electrical tape around three-quarters of the bottom to act as a spacer against the wood.

Measure Results

Take readings at the same time every day, or every several days, if precipitation is infrequent. If you collect snow, bring the jar inside and allow the snow to melt before making your measurement. Five inches (12.5 cm) of snow melts into only ½ inch (1.25 cm) of water. Add a precipitation column to your chart to record the data.

Thermometer & Box

You Will Need

- Mountable outdoor thermometer
- 2 small window-shutter panels
- 2 squares of ½-inch (1.25-cm) thick plywood (one slightly larger than the other)
- Rectangle of ½-inch (1.25-cm) thick plywood for the back
- 1 × 1 (2.5 × 2.5 cm) wooden post 5½ feet (165 cm) long
- Medium nails
- Long nail
- Small screw
- Protractor

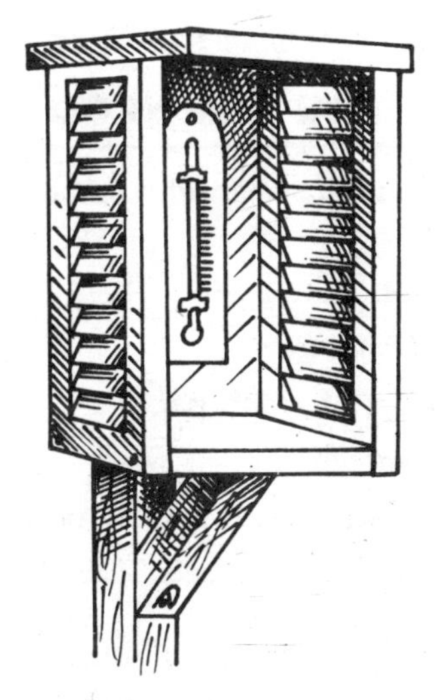

Thermometer & Box

Thermometer

Thermometers need protection from the elements. Here shutters provide shelter from hot sun and allow for ventilation—important for an accurate air-temperature reading.

Box dimensions can vary according to the size of your thermometer and shutter panels. Use the smaller plywood piece for the base, making sure it's wide enough to fit cleanly inside the shutters. Center the larger roof piece at the top of the box so that it creates an overhang; then record the position with a pencil mark on the wood.

Assemble the box by first nailing the shutters into the base, then nailing the roof into the shutters. Check to make sure the structure is squared and steady. Then nail the rectangular back panel to the box.

Construct the box mounting by sawing off a 6-inch (15-cm) piece from the 1 × 1 post. Place the piece horizontally before you, and, using the protractor and pencil, draw a 45-degree line from each bottom corner. Saw along these lines, securing the piece in a table vise if necessary. This piece will be attached to the top of the mounting post and support the box from below.

Continue using the saw and vise to shave the bottom of the longer post into a point. If this is too difficult, dig a hole in the ground to secure the post in place.

To attach the box to the mounting, center the long post against the box's base along the back edge. Hammer a nail through the base into the top of the post. Position the smaller piece so that the angled cuts are flush against the long post and base of the box. Hammer nails through the ends of the piece.

You don't need an expensive outdoor thermometer, just something with large, clearly marked numbers and a screwhole at the top for easy mounting. Turn the box so that the open side faces you, and screw the thermometer to the back panel, centering it as best you can.

Set up the box facing north. Record your temperature reading at the same time every day.

Barometer

You Will Need

- Medium glass jar with small mouth
- Balloon
- Drinking straw
- 4 pieces of graph paper
- Rubber cement
- Red marker with narrow point

Su M T W Th F S

Barometer

Air Pressure

Barometers measure air pressure and provide information about changing weather patterns. A **high-pressure system** usually indicates fair weather without precipitation, and a **low-pressure system** suggests warmer clothing and possible precipitation. Data you collect from a barometer may be the first warning of cooler temperatures, increased wind, and precipitation to come.

Barometer Construction

Flatten the straw 1 inch (2.5 cm) from the end. Then, cut diagonally from the corner of the flattened end to create a sharp, quill-like point. Color the tip with a red marker.

Cut the balloon in half and stretch the balloon across the jar's mouth, securing it with a rubber band to create a tight, drum-like surface. Apply a strip of rubber cement along the stretched balloon from the center of the bottle's mouth to the edge, and carefully place the unflattened end of the straw there, holding it in place until the glue dries. The straw should stick out about 6 inches (15 cm) perpendicularly from the bottle.

Recording Data

Divide each piece of graph paper into seven columns, one for each day of the week. To keep your data accurate, tape the paper to a wall and place the bottle close by on a flat tabletop. Now position the bottle in front of the first column, say, Sunday, so that the straw's red point almost touches the graph paper. At the time of day you make your regular temperature, wind, and precipitation measurements, note the position of the pointer and make a small pencil mark on graph paper. Then, move the bottle to the next column. Repeat this procedure each day for the duration of your weather window.

Barometer Readings

Watch your barometer work each day. When snow or rain is likely, the air inside the bottle will be heavier than the air outside the bottle, and so the balloon will stretch out a little, dipping the pointer down. A nice day means the air outside the bottle will be heavier than the air inside the bottle and the balloon will press down into the bottle, moving the pointer up. Each day, mark the pointer's position on graph paper. After the first week, remove the graph paper and replace it with a fresh piece. Connect the dots with lines to create a picture of air-pressure patterns.

Weather Vane & Wind Gauge

You Will Need

- 9 × 11-inch (22.5 × 27.5-cm) sheet of galvanized tin
- 21-inch (52.5-cm) stiff wire or wire coat hanger
- Steel supporting rod ¼ × 18 inches (.62 × 45 cm)
- ¾-inch (1.9-cm) pine board 5 inches (12.5 cm) square
- ⅜-inch (.95-cm) thick pine board 7 × 8¾ inches (17.5 × 21.8 cm)
- Wood strip 1¼ × 7 × ¾ inches (3.1 × 17.5 × 1.9 cm)
- Wood strip 4½ × ¾ × ¾ inches (11.25 × 1.9 × 1.9 cm)
- 4 small corner braces with screws
- 1 × 1 wooden post 6 feet (180 cm) long
- 11 small screws
- 2 small nails
- Two ½-inch washers
- 2-hole set-screw coupler
- ⅜-inch screw eye
- 7½-inch (18.75-cm) string
- Tin snips
- Drill
- Soldering iron
- Pliers with wire cutters
- Outdoor paint in two colors, light and dark

A Matter of Direction

Weather forecasters describe the wind by the direction it comes from, not by the direction it goes. So, a northeasterly gale originates in the northeast and pushes you in the opposite direction. Good weather vane design takes this into account. The narrow, pointed end indicates the wind's direction of origin while the wider side acts as a sail.

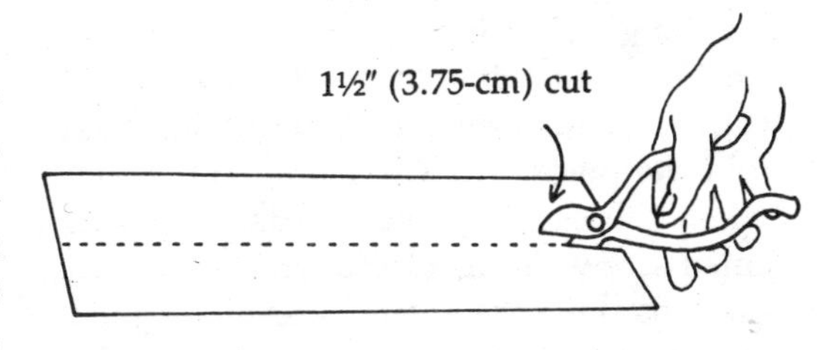

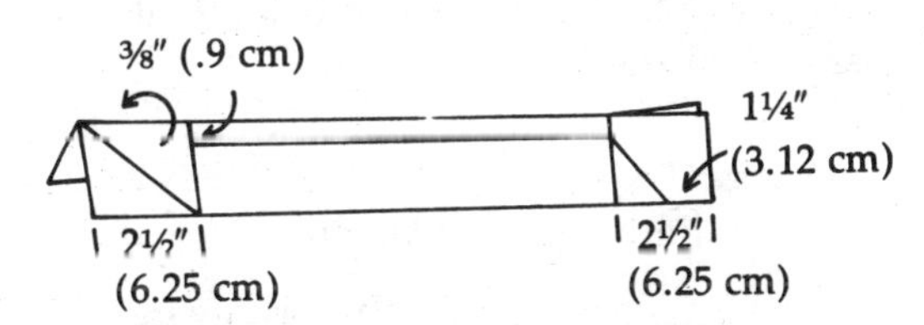

Weather-Vane Construction

Constructing the Tin Arrow

Fold the 3 × 11-inch (7.5 × 27.5-cm) tin strip in half horizontally. Draw an arrow and cut it out with tin snips. (See diagram.)

Constructing a Radial Indicator & Pendulum Disk

Place the 7 × 8¾-inch (17.5 × 21.8-cm) piece of pine horizontally in front of you. Starting from the top right corner, measure down the right side 3 inches (7.5 cm) and mark it. Moving right to left from the mark, draw a perpendicular line 2⅜ inches (6 cm) long.

Connect the end of that line to the bottom

right corner with a quarter-circle having a 7-inch (17.5-cm) radius. To do this, tie the pencil to one end of the 7½-inch (18.75-cm) piece of string. At the top left corner, secure the free end of string with your thumb, and sweep the pencil around in an arc.

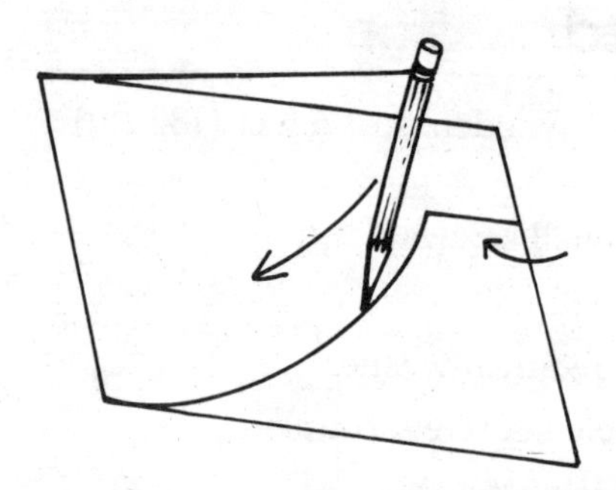

Radial Indicator

Use the same technique for drawing the pendulum disk on the 5-inch (12.5-cm) square pine. First, bisect the wood vertically and horizontally to find the exact center. Shorten the string to 2½ inches (6.25 cm), and hold the free end at the center while sweeping the pencil around in a full circle.

Cut both the indicator and disk from the wood. Cut a notch in the disk 1 inch (2.5 cm) long and ½ inch (1.25 cm) deep. Using tin snips, cut a double-sided arrow from the extra tin, drill a small hole at the center, and fold the piece into a U shape so that it fits neatly inside the notch. Attach it with a small screw. Next, carefully drill two small holes flanking the notch, close alongside the arrows. Cut a 1-inch (2.5-cm) piece from the 21-inch (52.5-cm) wire, and bend it into an L shape for the lock wire.

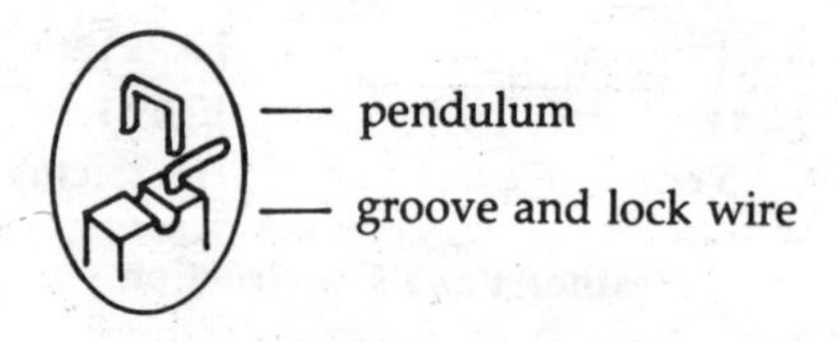

Complete the pendulum by bending the remaining 20 inches (50 cm) of wire into a narrow U shape and inserting the ends into the holes on the disk. Make sure the wire sits snugly inside the holes, and that the disk is not in danger of slipping off. Cut a small ¾-inch (1.9-cm) groove at the top of the wooden strip to cradle the pendulum. Alongside the groove, drill a hole to contain the lock wire.

Putting It All Together

Attach the wooden strip to the radial indicator's front with the small corner irons, two on each side. Make sure you attach the strip with the groove perpendicular to the indicator.

Also attach the tin arrow to the indicator, screwing the flared ends to opposite sides. Be careful to leave just enough room in front of the strip for the ¼-inch (.62-cm) steel support rod. At the strip's bottom, attach the screw eye.

Screw the end plates to the back of the indicator and flare them outward with the 4½-inch (11.25-cm) strip of wood, nailed into place. Hang the pendulum from the notch and close the lock wire. Make sure the pendulum swings freely.

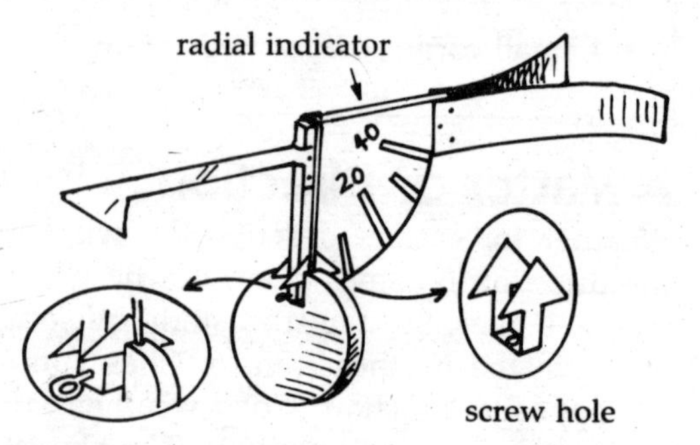

Weather Vane & Wind Gauge Assembly

Going for a Ride

Calibrate your wind gauge so that you can paint radial lines indicating wind speed. Have someone take you for a car ride, while you hold your weather vane out the window and observe the pendulum's position at various speeds. With your pencil, make a hatch mark along the indicator's curved edge at 10, 20, 30, and 40 miles (or km) per hour.

Paint & Finish

Paint the indicator's background a dark color and the radial lines a light color. Paint both sides of the indicator.

After you mount your instrument on a steel support rod, you can begin collecting data. Measure 7½ inches (18.75 cm) from the top of the rod and attach the set-screw coupler, tightening both screws until the coupler grips firmly. Slip the two ½-inch washers over the rod until they catch on the coupler. Make sure they sit level.

Slide the weather vane over the rod's top—first fitting the rod through the eye-screw, then through the space where the tin arrow attaches to the wooden strip. The instrument should rest on the two washers, supported below by the coupler.

Plant your weather vane in the ground. Keep a compass close to determine the wind's direction as the weather vane shifts. Take wind direction and velocity readings at the same time each day, entering data in your notebook.

Weather Station

You can create your own weather station by constructing all five projects—Thermometer & Box, Sling Hygrometer, Weather Vane & Wind Gauge, Barometer, and Rain Gauge. Make a chart on a large piece of poster board displaying all your weather data. You could include date and time, general weather conditions (fair, cloudy, etc.), temperature, humidity (percentage), wind direction and speed (e.g., NE/15 mph), precipitation (None or Yes), and atmospheric pressure (High or Low).

Finding Mean Astronomical Time

You Will Need

- World map or globe
- Notebook

A Matter of Timing

For most of us, the time we read on our wristwatches, **standard civil time**, does not accurately reflect the real time of astronomical events. In fact, a sundial gives us a more accurate reading because it represents **mean astronomical time**—time measured by the movement of the sun, moon, and planets across the sky. Mean astronomical time may differ from standard time by many minutes, depending on the observer's location.

Since the earth is round, when you travel the ground curves slightly and a different part of the sky appears directly overhead. When the sun is at its zenith in one location—**local apparent noon**—it's a minute short of noon a few miles west, and a minute past noon a few miles east.

Odd Times

Only a little over a century ago, cities and town across America set their clocks slightly differently to reflect these variations. As railway systems developed, confusion resulted from the many different local times used along the lines. The first step towards developing standard time was the introduction of so-called **railway time**—the local civil time of an important rail station. As interstate and international communication expanded with the invention of the telegraph and telephone, people all over the world saw the need for a universal standard.

What's Standard

In 1884, the International Meridian Conference adopted a system of standard time, dividing the earth's surface into 24 zones. The standard time for each zone is the mean astronomical time of one of 24 **meridians** (lines of longitude), 15 degrees apart, beginning at Greenwich, England. These meridians extend east and west around the globe to the **international date line.**

Standard time is the mean astronomical time at an agreed-upon standard longitude. In the continental United States, for instance, these longitudes are 75 degrees (eastern standard time or EST), 90 degrees (central time), 105 degrees (mountain time), and 120 degrees (Pacific time).

For practical purposes, this convention is sometimes altered. Some smaller countries, for instance, still use the local civil times of their individual capitals as standard everywhere within their borders.

Mean Astronomical Time Calculations

If you live directly along one of the meridians, standard time and mean astronomical time are the same. But if you live between meridians, you can easily calculate mean astronomical time using only a map, notebook, and pencil to do simple arithmetic.

Consult the map to determine how many degrees of longitude you are from your time-zone standard. Then multiply this number by 4 to find your correction in minutes. If you're east of standard longitude, your correction is a *positive* number; if you're west, it's *negative*. Apply the correction to standard time to get your mean astronomical time. Subtract an hour if daylight saving time (DST) is in effect.

For example, Cambridge, Massachusetts, sits at 71 degrees west longitude. Subtract this from 75 degrees (since eastern standard time applies here) and multiply by 4. The correction is *plus* 16 minutes. So 12:00 midnight eastern standard time is 12:16 A.M. mean astronomical time. This means that in Cambridge, celestial objects rise, cross, and set 16 minutes *earlier* than they do at the standard longitude.

Compare your local mean astronomical time with mean astronomical time at various global locations. Where would you find the greatest difference between the two measurements? List your results on a chart.

Moon Craters & Martian Channels

You Will Need

- 2 shallow cardboard boxes (shirt boxes) about 3 × 3 feet (7.5 × 7.5 cm) and 2 to 3 inches (5 to 7.5 cm)
- Piece of plywood 3 feet (7.5 cm) square
- Aluminum foil
- Plaster of Paris
- Sand
- Sugar
- 2 disposable paint-mixing pails
- Wooden paint stirrer
- Wooden yardstick
- Small bowl
- 2 saucers
- Small stones and pebbles, aquarium gravel, and sand
- 1 foot (2.5 cm) of narrow rubber tubing
- Watering can
- Paper cup
- Carpenter's glue
- 2 shades of gray spray paint
- 2 shades of yellow or brown spray paint
- Spray varnish or polyurethane

Moon Surface

Though "dead"—waterless, lifeless, and without the barest hint of an atmosphere—the surface of our closest neighbor nonetheless bristles with fascinating detail. Only recently, however, have scientists begun to understand the forces behind its amazing geological features.

The **craters** of the moon, or of any other planet, were formed in one of two ways: by meteors colliding with a molten surface or by erupting volcanoes. In this project, you can easily simulate both activities and create, in the process, a realistic and permanent craterscape.

Creating a Craterscape

Secure the sides and corners of the boxes with masking tape, and line the insides with aluminum foil. Place one of the boxes on the floor.

On a tabletop close by, lay out a collection of stones, pebbles, and marbles. Also place a sand-filled saucer, a gravel-filled saucer, and a bowl filled with water on the tabletop.

Mix enough plaster in the paint-mixing pail to completely fill the box. If you add about ½ cup (120 ml) of sugar to the plaster mixture, the plaster will take longer to harden. Mix the plaster to a putty-like consistency. Using the wooden yardstick as a trowel, scrape off the excess plaster from the top of the box until you have a smooth surface.

Now you can begin to make craters. Choose among the various sizes of stones and pebbles, and simply drop a few into the plaster. Next, back up and *throw* a stone and some pebbles, varying the force and angle of your throw. Do the same with a pinchful of sand, wet and dry, and some gravel, but avoid covering the surface with too much detail just yet.

Next, crouch down over the tray, and maneuver one end of the rubber tube through the plaster so that it remains buried, but points towards the surface. Blow through the other end of the tube to simulate a volcanically created crater. Repeat the process in several places.

Continue making craters and volcanoes until the plaster hardens and it no longer retains impact impressions.

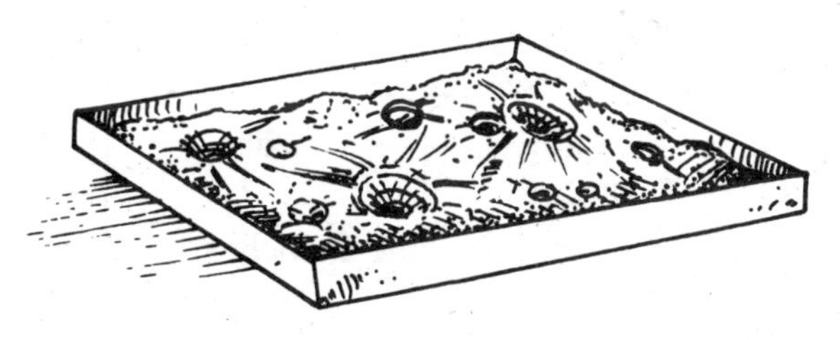

Craterscape

Allow the craterscape to harden overnight before you attempt to remove it from the box. Then remove it by breaking away the sides of the box and peeling off the aluminum foil.

Finishing the Craterscape

The final step consists of spray painting the model, first with the lighter of the two shades of gray. Make sure you thoroughly cover the entire surface and allow it to dry. The darker gray provides the finishing touch and a dramatic effect to your craterscape. Hold the spray can at about 45 degrees to create shadows along the far side of your craters. These shadows will highlight the impact detail of each crater as well as the many other geological nuances of your simulated lunar terrain.

Martianscape

For the surface of Mars, we will concentrate on reproducing Martian channels—geological formations that resemble the dry riverbeds of earth. Indeed, scientists now agree that there was once water flowing across the surface of the red planet, and the erosion of sand from the force of that water created the channels.

Place the second box on the plywood. Mix a little water with sand in the pail so that the sand clumps together slightly. Pack the moist sand into the box, filling it to the top. Use the ruler to smooth the surface, and place a few stones here and there.

Move the box outdoors where you can freely douse it with water. Incline one side to about 45 degrees, either by propping it with a few bricks or by having someone hold it for you. Punch a few holes in the bottom of the paper cup with a small nail. Fill the watering can.

Begin the erosion process by pouring a stream of water from the can against the inclined side of the tray. Move the stream back and forth so that erosion occurs along the entire plane, particularly alongside the rocks. If someone holds the box for you, the helper could vary the degree of inclination as you continue to pour.

Next, add water to the paper cup so that it showers through the perforations. Suspend the shower against the inclined side of the tray for a more delicate erosion pattern.

Place the box flat, and allow several days outdoors for the sand to dry. Or, you can use a hair-dryer to remove moisture from the sand, taking care to keep the flow of air far enough away from the surface to prevent damage to the patterns.

When the sand dries thoroughly, apply several coats of spray varnish, allowing several hours between each coat. When the surface appears fixed and hardened by the varnish, carefully break the box apart, and peel the foil from the sand. Finally, scrape the loose sand from under the varnish-hardened surface, and mount your Martian landscape on the plywood with carpenter's glue. As with the craterscape, use the lighter shade of paint first, followed by the darker shade to bring out surface details.

Display your models side by side with an explanation of how each was formed.

Orrery

You Will Need

For the Post

- Threaded hollow lamp rod 22 × ¼ inches (55 × .62 cm)
- Copper pipe 18 × 1 inches (45 × 2.5 cm)
- Copper tubing ½ × 5⁄16 inches (1.25 × .79 cm)
- ½-inch (1.25-cm) plywood base 12 inches (30 cm) square
- 2 plywood disks ½ × 3½ inches (1.25 × 8.75 cm)
- ¼-inch (.62-cm) - thick plywood disk 6 inches (15 cm)
- 2 ball-bearing turntables 3-inch (7.5-cm) diameter
- Screw-in lamp socket with electrical cord
- 40-watt light bulb (clear)
- 4 screw-in furniture glides

For the Pulley & Arm

- Plywood disk ½ × 2½ inches (1.25 × 6.25 cm)
- Plexiglass tube ⅞ × 3½ (2.22 × 8.75 cm)
- Ball-bearing sliding glass-door roller ¼ × 1 inch (.62 × 2.5 cm)
- 2 plastic foam snow balls, baseball and Ping-Pong sizes
- Stiff wire or coat hanger
- 2 wooden dowels 14 × ¼ inches (35 × .62 cm)
- Cotton wrapping twine 44 inches (110 cm)

Additional Materials

- Drill with ¼-inch and 5⁄16-inch bit
- Machine bolt ¼ × 1½ inches
- 4 fender washers
- 3 lock washers
- 2 rubber washers
- Three ¼-inch nuts
- Two ¼-inch wing nuts
- Table vise
- File
- Medium-gauge sandpaper
- 8 small wood screws
- Craft knife
- Electrician's tape
- Epoxy glue

Planet Modelling

When the theories of Copernicus became widely accepted, astronomers began building models to demonstrate the movement of the earth and moon around the sun. The models were expensive, and some were more accurate than others. But the industry of astronomical model building grew, and like the industry of clockmaking, it produced some gorgeous, if not fanciful, creations.

The most famous of these was a device owned by the English nobleman Charles Boyle (1676–1731), the 4th Earl of Orrery. His "mechanical plannet clocke," still on display in a British museum, was a wonder

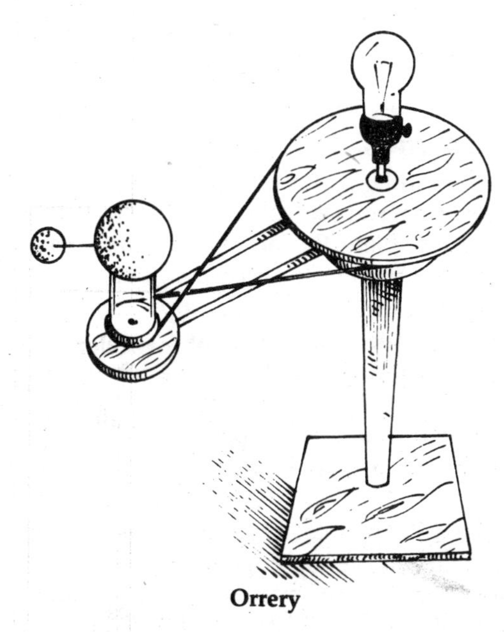

Orrery

in its time because it showed the earth, moon, and sun all revolving at the correct speeds relative to each other.

You can design a much simpler version of Boyle's "orrery," as it came to be known. But your model can nonetheless demonstrate important astronomical principles.

Post Parts

The orrery described here is basically a modified lamppost with a swinging-arm attachment. Begin with the base. Divide the 12-inch (30-cm) square plywood in half both vertically and horizontally. Drill a ¼-inch (.62-cm) hole in the center, where the lines intersect. Screw in the furniture glides at the corners of the plywood.

Place the two 3½-inch (8.75-cm) disks and the 6-inch (15-cm) disk on a flat surface in front of you. As with the base, determine the exact center of each disk by drawing a diameter line across the middle, then bisecting it with a perpendicular line the same length. Drill a ¼-inch (.62-cm) hole through the center of the 6-inch (15-cm) disk and through one of the 3½-inch (8.75-cm) disks. Drill a 5/16-inch (.79-cm) hole through the center of the other 3½-inch (8.75-cm) disk. Put the 6-inch (15-cm) disk into the table vise. Using the file, cut a straight groove completely around the edge. The groove shouldn't be too deep, just deep enough to cradle the cotton twine. When finished, remove the disk from the vise.

Take the 3½-inch (8.75-cm) disk with the larger hole, and measure ¾ inches (1.87 cm) from the center, on opposite sides of a diameter line. From the ends of these marks, draw perpendicular lines out and across to the disk's edge. Place the disk in the table vise with the marked edges facing you, and using the ¼-inch bit, drill holes over the marks. You want parallel bores ½-inch (1.25-cm) deep, so make sure you drill straight, and not towards the disk's center.

Remove this disk from the vise, and hammer the ½-inch (1.25-cm) long piece of copper tubing through the center hole. On both sides of the disk, center the ball-bearing turntables. Screw only one plate of each turntable to the disk, leaving the other plate free.

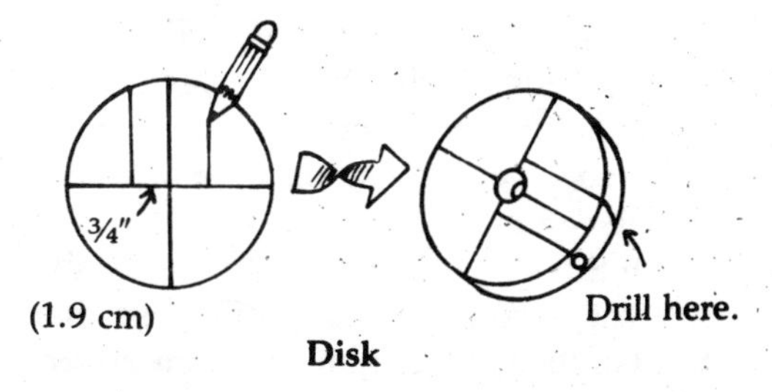

Disk

Trace the other 3½-inch (8.75-cm) disk onto a sheet of sandpaper. Cut out the sandpaper circle and glue it to the disk.

Assembling the Post

Insert the 22-inch (55-cm) hollow rod through the 18-inch (45-cm) copper pipe. Attach the base to one end, securing it with a fender washer and wing nut. Stand this upright and push the sandpaper-bottomed 3½-inch (8.75-cm) disk, paper side down, over the rod until it's flush with the top of the pipe. You may need to gently hammer the disk down along the rod. To make sure the disk grips the top of the pipe tightly, place a fender and lock washer over the rod and tighten them with a nut.

Next, slip the disk with ball-bearing turntables over the rod until it sits level with the first disk. The copper tubing in the center hole should allow the disk to slip smoothly down the rod. Finally, place the 6-inch (15-cm) grooved disk over the rod, tightening it with a fender washer, lock washer, and nut. You should have a sandwich of three disks, with the middle disk freely rotating between the upper and lower stationary disks.

Assembling the Pulley & Swinging Arm

Find the center of the 2½-inch (6.25-cm) disk and drill a ¼-inch (.62-cm) hole there. Measure ¾ inches (1.9 cm) from the center, on opposite sides of a diameter line. From the ends of these marks, draw perpendicular lines out and across to the edge of the disk. Place the disk in the table vise with the marked edges facing you, and drill ¼-inch (.62-cm) holes over the marks. Again, you want parallel bores, ½ inch (1.25 cm) deep; so, make sure you drill straight, and not towards the center of the disk.

Attach the sliding glass-door roller to the disk with the ¼-inch machine bolt, so that the head of the bolt rests on top of the roller. Use two rubber washers between the roller and the disk. Tighten the roller to the disk with a wing nut and fender washer. Apply a small amount of epoxy glue to one end of the 3½ × ⅞-inch (8.75 × 2.22-cm) plexiglass tube, and press the glued end over the roller.

To attach the pulley to the arm and the arm to the post, apply a little carpenter's glue to one end of each dowel and insert those ends into the side holes of the 2½-inch (6.25-cm) disk. Apply glue to the other ends of the dowels, and insert them into the side holes of the post's middle disk. Allow the glue to dry.

Tie the ends of the cotton twine together so that you have a loop 21½ inches (53.75 cm) long. Cut a 2-inch (5-cm) section of twine from the loop, and replace it with a 1½-inch (3.75-cm) section of rubber band. Twist one end of the loop so that you have a figure *8*. Then slip one end over the grooved disk and the other end over the sliding glass-door roller. The twine should stretch tightly against both grooved disk and roller.

Earth, Moon & Sun

You need only to screw the lamp socket to the top of the rod and attach the plastic foam balls to complete the orrery. But first, you need to cut the electrical cord so that it can be threaded through the rod, top to bottom.

Cut near the plug. After pulling the wire through, screw the socket to the top of the rod, as far down as it will go. Place the 40-watt bulb in the socket. Splice the wire back again, using a craft knife and electrician's tape.

Glue the larger of the 2 plastic foam balls (the earth) on the top edge of the plexiglass tube so that the ball sits level with the light bulb. Cut 5 inches (12.5 cm) of wire or coat hanger and insert it into the side of the earth. At the other end of the wire, attach the smaller plastic foam ball (the moon). You might paint the earth blue, adding continents, and the moon, gray. Drawing a face on the side of the moon opposite earth helps illustrate how we always see the same side of our lunar neighbor.

Some Astronomical Demonstrations

Your orrery easily demonstrates basic astronomical principles. As you rotate the arm in a counterclockwise direction, notice how the earth moves around the sun in a counterclockwise orbit, just as the moon moves around the earth in a counterclockwise orbit. Notice, too, that the earth rotates 180 degrees on its axis—the length of one day—many times as it orbits around the sun, while the moon rotates only once on its axis as it travels around the earth. This is why we always see the same side of the moon and why we know so little about its dark side.

It takes the earth 1 year to orbit the sun, and during this time many days and nights pass as the earth revolves around its axis. Our orrery can only approximate the velocity of these revolutions. The difference in diameters between the disk and pulley creates a 6-to-1 relationship: six complete earth revolutions per solar orbit. Therefore our model shows us that each earthly revolution stands for six times that amount, or that each day represents 60 days.

The moon takes 27 days, or 1 **sidereal month,** to complete an orbit around the earth. Although this is not represented by the model, you can still determine phases of the moon and eclipses by studying the moon's position relative to the sun and earth.

The completion of the sidereal month is marked by the **new moon**—or a moon that doesn't appear at all during the night because it's situated between the earth and the sun. During this new moon phase, solar eclipses may occur. Continue to swing the arm around until the moon lies directly between the sun and earth, throwing its shadow back against earth. You now have a new moon and solar eclipse. A person standing in the moon's shadow on the daylight side of earth sees the eclipse; a person standing on the nighttime side of earth sees no moon at all.

Swing the arm slowly until the moon revolves 180 degrees and into the earth's shadow. When earth and moon align precisely, a lunar eclipse occurs and we see the shadow of our own planet creep across the moon's surface. Of course, actual orbits are elliptical rather than round, and tilted rather than flat, so an exact alignment between the earth and the moon do not occur frequently.

By continuing to swivel the arm while watching the moon from the vantage point of a person on the night side of earth, you can see the moon's phases replicated perfectly. The sidereal month (from new moon to new moon) is divided into four quarters during which the moon grows, then shrinks. The first quarter begins with the new moon, which first appears as a crescent moon and grows, or **waxes,** into a half moon. During the second quarter, the half moon waxes into a full moon, passing through a gibbous (three-quarters) stage. The full moon shrinks, or **wanes,** to a half-moon during the third quarter. And the half moon becomes a new moon again at the end of the fourth quarter.

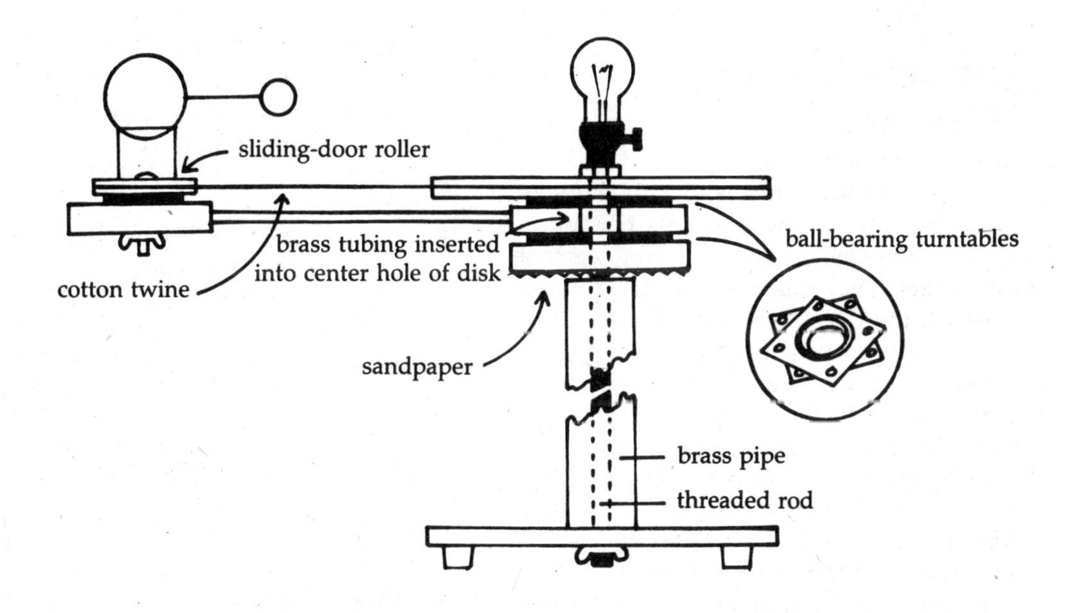

Orrery Assembly

Connect the Star Dots

You Will Need

- Large, stiff poster board in a dark color
- 2 sheets of medium blue transparent Mylar 9 × 12 inches (22.5 × 30 cm)
- Masking tape
- Cellophane tape
- Craft knife
- Metal ruler
- Hole punch
- Black construction paper 8½ × 11 inches (21.25 × 27.5 cm)
- White photocopy or similar paper 8½ × 11 inches (21.25 × 27.5 cm)
- Light-color pencil
- Low-voltage desk lamp, with rotating hood (if possible)

Constellations of "Fixed Stars"

Ancient astronomers once believed the stars were attached to the inside of a huge dome. They also noticed how other celestial bodies appear unattached and wander against these stars. They called a wandering object **a planet**, the Greek word for "traveller."

But the fixed positions of stars still fascinated the ancients, and they soon began connecting points of starlight to form pictures. These pictures, what we call **constellations**, differ from culture to culture. Where an ancient Chinese astronomer saw a dragon, a European saw a plow, and where an ancient Native American recognized a bear, an Arabian traced a jewelled robe. Eventually, these cultures recognized and accepted popular names for constellations.

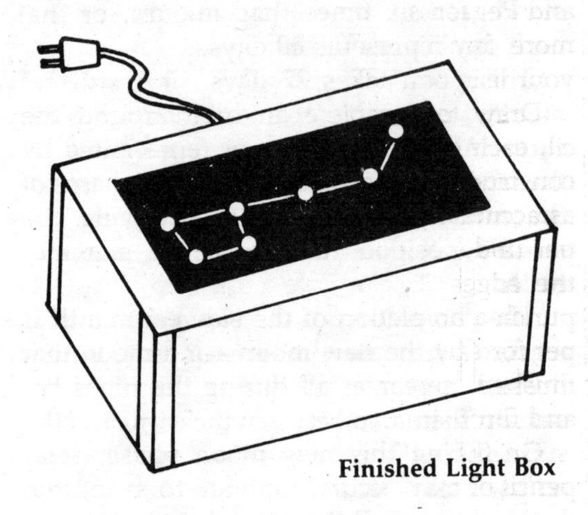

Finished Light Box

Light Box Construction

You can learn to recognize constellations with the aid of a simple light box and transparent slides.

Cut the poster board into the pattern, and assemble the box shown. Use masking tape along the vertical edges to strengthen the box; then, flip it upside down and put it aside.

Carefully place the two sheets of blue Mylar together so that the edges line up. Strip the cellophane tape around three of four edges, leaving one 12-inch (30-cm) edge untaped. You now have a Mylar envelope.

Position the Mylar envelope over the window cut into the poster board, with the untaped edge facing the rectangular opening in the back wall. Center the envelope over the window before attaching it to the poster board with strips of masking tape over the cellophane-taped edges. Tape carefully so that nothing obscures the window. Flip the box over and check the finished window to make sure the edges look clean and even.

Preparing Slides

Now you need to make the slides. Choose several easily recognized constellations and several less familiar ones, consulting a star chart. You could start with the Big Dipper, followed by Auriga, Cepheus, and Perseus; then Libra and Hercules, followed by Draco and Pegasus. List your choices, placing the more familiar constellations at the top of your list.

Draw, using a ruler and light-color pencil, each constellation on the black piece of construction paper, reproducing the shape as accurately as you can. Place a white paper under the construction paper, aligning the edges. Use a hole punch to carefully punch a hole through the two pieces of paper for each star in the constellation. When finished, separate the two pieces of paper and flip them over.

On the white paper, draw, with a dark pencil or marking pen, the lines connecting these "stars." Write the constellation's name as well as any other useful information at the bottom. Place the black construction paper over the white paper so that the holes line up. Clip them together with a paper clip until show time.

Star Dots

To use the light box and slides, first place the lamp under the box, rotating the hood so that the bulb radiates up towards the blue Mylar screen. Remove the clip from the sheets of paper while carefully holding them together. Then slide them, black construction paper side up, through the back opening and between the two sheets of Mylar. Blue light will poke through the holes, outlining the constellation. However, without the connecting lines the constellation may be difficult to recognize.

Challenge viewers to identify the constellation. Then reveal the answer by connecting the star dots—gently sliding the construction paper out of the Mylar envelope, revealing the white paper underneath.

Or give viewers copies of disconnected star dots on separate sheets of paper and ask them to draw in the constellation. You could provide a few cues, such as "This group includes the North Star" or "This has a zodiac name."

You could include more elaborate drawings over the simple connected dots to illustrate, for example, the crab of Cancer or the giant Hercules. Consult star maps for this, especially reproductions of fanciful star maps used by ancient astronomers.

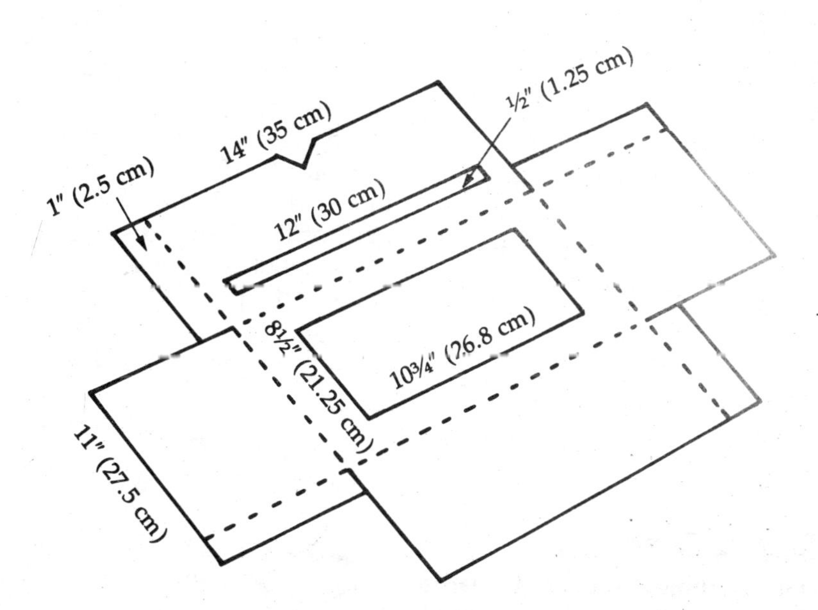

Pattern for Light Box

Crystal Planetarium

You Will Need

- Chemist's flask with round bottom at least 6-inches (15-cm) diameter
- Rubber stopper for flask
- Glass or plexiglass rod
- Bowl
- Small piece of ⅛-inch (.31-cm) - balsa wood
- Dark blue ink *or* blue food coloring
- Phosphorescent paint
- Narrow-point paintbrush
- Ink eradicator (white)
- Wineglass
- 2 plastic pushpins
- Protractor
- Craft knife
- Grease pencil
- Red and blue rubber bands

Crystal Planetarium

Crystal Gazing

You can easily duplicate a large planetarium's workings with this small model, constructed from an ordinary round-bottom chemist's flask. Gazing through your crystal planetarium in a dark room, you'll see a miniature spectacle of constellations rising above and setting below the horizon of a becalmed ocean. And in the middle of it all, a tiny steamer makes its way to some mysterious destination.

Stars, Sea & Steamer

Drill a hole in the rubber stopper, just large enough for the glass rod. The rod should reach the flask's bottom when it's corked tightly. Next, remove the cork and rod, and fill the flask with enough clear water so that the flask is half full when turned upside down. Use your hand to stop the opening while determining this water level. Finally, pour the water into a separate bowl, and color it deep blue with food coloring or ink.

Using the craft knife, construct the steamer from two pieces of balsa wood glued together. Two plastic push-pins will work for the funnels. Just make sure your model is small enough to fit through the neck of the flask. To stabilize the steamer, make a slot underneath with a craft knife and insert a penny. Float the steamer in the bowl of blue water to check its stability, making any necessary adjustments.

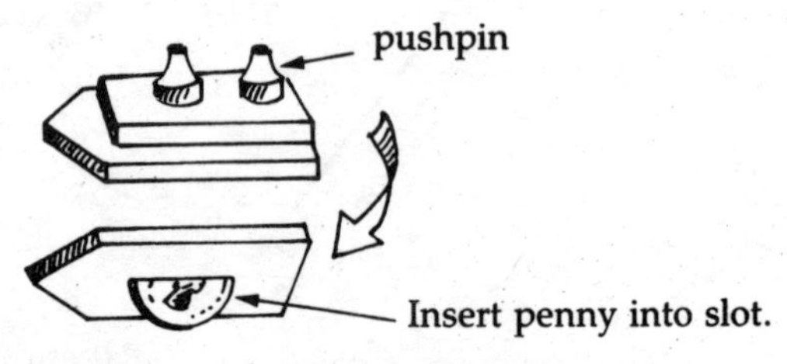

Balsa-Wood Boat

Preparing the Flask

To transform your flask into a planetarium dome, place the red rubber band around the middle of the flask for the equator. Next, place the blue rubber band at about a 23-degree angle to the equator (use a protractor). The blue band represents the **ecliptic**—the path of the planets, moon, and sun.

Using a tape measure, divide the distance from the equator to the poles into equal thirds. Each third stands for 30 degrees, and the whole quarter-circle shown in the illustration for 90 degrees. Mark the divisions with grease pencil circles extending around the flask, parallel to the equator.

With these lines to guide you, transfer the quarter-sphere maps below onto the surface of the flask with lightly applied grease pencil. Since these "stars" are supposed to be viewed by people in your miniature steamer inside the flask, star group figures are reversed on the maps.

When you're satisfied with your maps, remove the rubber bands and go over each grease pencil mark with a dab of phosphorescent paint. In some cases, you may want to use a cloth moistened with alcohol to remove the grease pencil dot before replacing it with paint. (Paint one side of a small coin—our sun and moon for future demonstrations.)

After the paint dries, use a dab of ink eradicator (white) on each star dot. This ensures that your stars glow only on one side.

Putting It All Together

Replace the equator and ecliptic (red and blue rubber bands). Carefully pour the blue water into the flask and add the steamer. Seal the flask with the rubber stopper and glass rod, and set your completed planetarium on the wineglass with the glass rod pointing north and meeting the water at an angle equal to your latitude. (New York, for instance, is about 40 degrees north latitude, Toronto is roughly 44 degrees north latitude, London is about 51 degrees north latitude, and Sydney is about 34 degrees south latitude.)

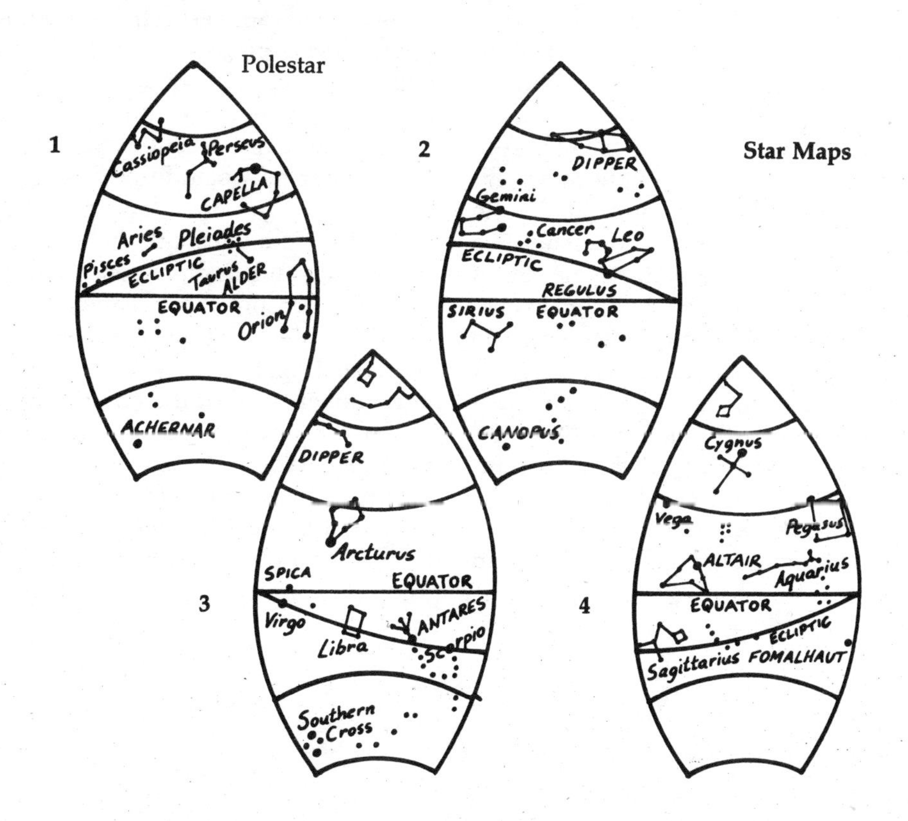

Star Show

Let's put the planetarium to work. Notice how the end of the glass rod points directly at the **polestar**. Dim lights, hold the flask's neck, and turn the flask slowly counter-clockwise. As you look through the flask's near side, some stars on the far side rise over the ocean horizon, cross the sky, and set on the left. Other **circumpolar stars** simply go around the upper end of the slanting glass rod, which represents the celestial globe's axis. These stars never set. Other stars, near the flask neck below the water surface, never rise in northern latitudes.

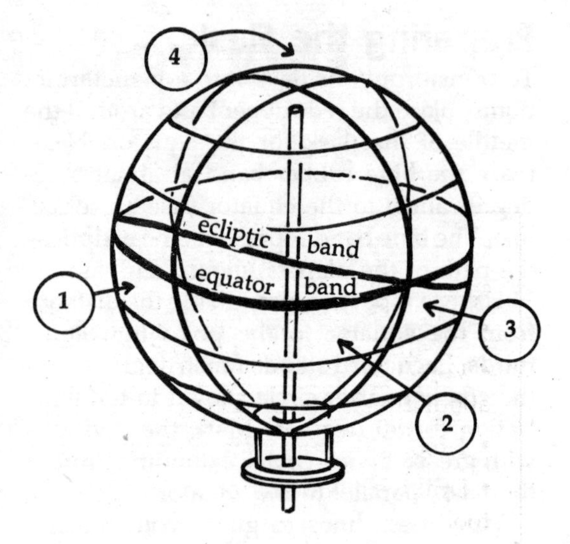

Placement of Star Maps

Polar Expedition

Now let's travel from pole to pole and compare differences in star maps. To go north from your latitude, remove the flask from the wineglass and gradually lower the neck. The glass rod representing the polar axis of the sky rises. When the rod is vertical, rotate the flask slowly between your hands to see the sky as it would appear if you stood at the North Pole. Straight overhead is the polestar, and the entire heavens revolve around it. The constellation Orion will always be on the horizon because the horizon remains parallel to the sky's equator.

Slip the painted coin under the blue rubber band representing the ecliptic at a point below the equator band, and rotate the flask. You'll see why the sun does not rise above the horizon during the 6 months it passes over half of its path that remains below the horizon at the North Pole. Now slip the dime under the ecliptic band at the other side where the band rises above the equator. When you rotate the flask, you'll see the midnight sun, never setting below the horizon for the six-month-long summer day.

Tip the flask back, place it on the wineglass, and continue raising the neck. When the glass rod is level with the water surface, the stars' position represents what we would see when our ship reaches the equator. The polestar is on the northern horizon, and if it's winter, Orion will pass through the zenith, directly overhead.

If you continue tipping the flask until the glass rod is again vertical, you'll see stars as South Pole explorers saw them. Orion, the equatorial constellation, stands upside down on the northern horizon. In fact, all the northern constellations visible south of the equator now stand upside down. Also, the midday sun slants down from the north instead of from the south. In a Buenos Aires hotel you would ask for a room with a sunny *northern* exposure. Your planetarium makes it easy to demonstrate this with the dime (representing the sun) slipped under the blue rubber band of the ecliptic.

If after your demonstration you want your planetarium displayed on a desk or tabletop, replace the wineglass with 8-inch (20-cm) transparent plexiglass tubing, cemented to a plywood base.

Umbrella Planetarium

You Will Need

- Large black (disposable) umbrella
- ¼-inch (.62-cm) wooden dowel about 2 feet (60 cm) long
- 1-foot (30-cm) square plywood
- Drill with wide bit or wood-boring tool
- ½-inch (1.25-cm) rubber or cork stopper
- Wire coat hanger
- Pliers with wire clippers
- White adhesive paper
- White or yellow grease pencil
- White chalk
- Hole punch
- Carpenter's glue
- Flat black spray paint
- Desk lamp with adjustable hood
- Punk *or* incense stick

Umbrella Planetarium

Clear Skies under Your Umbrella

You can make a planetarium out of a black umbrella. To make the adjustable umbrella stand, drill or bore a hole in the center of the plywood just large enough for the wooden dowel to fit snugly. Apply glue to the inside edge of the hole before inserting the dowel. Allow the glue to dry thoroughly before painting the stand black. Then, allow the paint to dry overnight.

Next, drill a hole in the center of the rubber or cork stopper. You want the stopper to fit tightly over the dowel, but be loose enough to slide along its length. Drill a smaller hole in the stopper's side to contain the hook. Make the hook from a wire coat hanger, snipping off a piece 1½-inch (3.25-cm) long and curling one end into a hook shape with pliers. Insert the straight end of the hook into the stopper.

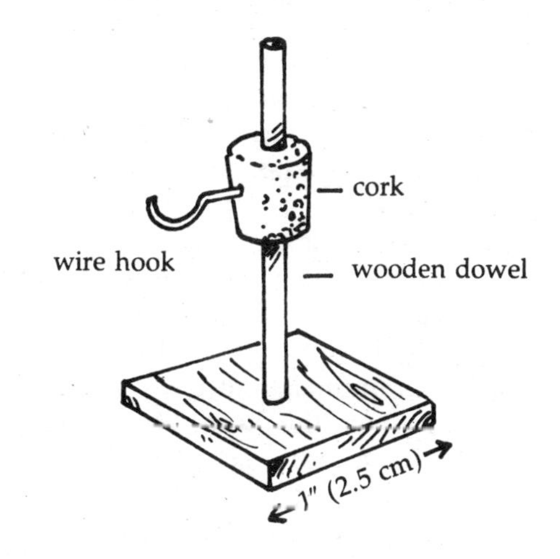

Umbrella Stand

Star Creation

Hold the adhesive paper over the saucer and punch about 100 holes through the paper. When the saucer contains enough "stars," discard the remaining paper. You may need a straight pin to separate the paper from the adhesive when you mount your stars on the inside of the open umbrella. You can substitute ink eradicator (white) for the adhesive stars, but this allows less room for error and adjustment.

Next, you want to recreate the constellations of the northern hemisphere on the umbrella's inside surface. But first, paint the metal ribs and pole of the umbrella black to hide them. After the paint dries, place the **North Star** (Polaris) next to where the umbrella pole joins the umbrella fabric—at the central axis around which all other stars revolve.

If you divide the umbrella into eight pie slices radiating from the North Star, you can align key stars of the major constellations with the North Star. Use the chalk to sketch positions before you affix the adhesive stars.

Start with the end star of W-shaped **Cassiopeia,** then proceed directly across from it to the central star of the **Big Dipper,** the star joining the dipper to the handle. Proceed clockwise to the **Little Dipper, Draco, Cepheus, Perseus, Auriga,** and **Cancer.** Then plot out all constellations in between. If your star map seems accurate, attach the adhesive stars. Finally, connect the stars of each constellation by drawing a dotted line between them with grease pencil. Label the constellations with the grease pencil, too, but avoid making the labels too large.

Loop the rubber band around the umbrella pole several times to prevent it from slipping as it leans against the stand. Then, point your umbrella north and incline it to a position where the North Star corresponds to the North Star's actual position in the night sky. You could take your umbrella planetarium outside at night to determine this angle precisely.

Working Planetarium

Place your planetarium on a table so that the umbrella part leans out over the edge. The table edge simulates the horizon, and if you place your eyes level with it while slowly revolving the umbrella counterclockwise, you can imitate the rising and setting constellations.

Notice how Cassiopeia and the Big Dipper circle around the North Star. These constellations do not rise and set because they are always seen above the horizon. Watch **Vega** set in the west below the horizon (tabletop) as you turn the umbrella, simulating the earth turning on its axis. Then, notice how Vega rises in the east as you continue to turn. Remember: the stars do not really move, they only appear to do so because the earth is rotating.

Light Effects

For a more dramatic effect, you could light your planetarium from behind, making holes in the umbrella for the constellations' positions. This requires more preparation.

To make a clean hole, use a burning punk or incense stick. Ask an adult to help, since the lit stick could cause injury. Touch the tip of the stick lightly against the umbrella's surface until you form a small, round hole. After marking out the constellations, spray the umbrella outside with black paint to make the fabric opaque. Allow it to dry overnight.

Next, place small squares of frosted cellophane tape on the outside of the umbrella over each hole. This is time-consuming, but it makes a more realistic starlight when the lamp shines through. Finally, position the lamp directly behind the umbrella, and drape off the area where your viewers will stand, waiting for the show.

Theodolite

You Will Need

- 6-inch (15-cm) plastic half-circle protractor
- 6-inch (15-cm) plastic full-circle protractor
- Plastic drinking straw
- Thick shirt cardboard 1 × 2 inches (2.5 × 5 cm)
- Wood for post, 12 × 1 × 1 inches (30 × 2.5 × 2.5 cm)
- Square plywood 8 × 8 inches (20 × 20 cm)
- 6-inch (15-cm) nylon string
- Small metal fishing weight
- 1-inch metal screw
- 1½-inch metal screw
- 2 washers
- Epoxy glue
- Flashlight
- Red cellophane

Sextant

Medieval astronomers and navigators used an early version of the sextant, the **theodolite,** to measure not only the altitude of a celestial object, measured from the horizon in degrees, but also its azimuth, or the horizontal distance from the North Star. Finding an object with a theodolite and recording the coordinates for a given date and time will help you locate the same object again.

Theodolite Construction

Cut the 1 × 2-inch (2.5 × 5-cm) cardboard square into an arrow shape, first dividing it in half lengthwise and measuring from the ends of the center line to the mid-point of each edge. Paint the arrow so that it sharply contrasts with the color of the full-circle protractor. Allow the paint to dry; then, glue the arrow to the bottom of the 1 × 1-inch (2.5 × 2.5-cm) upright post.

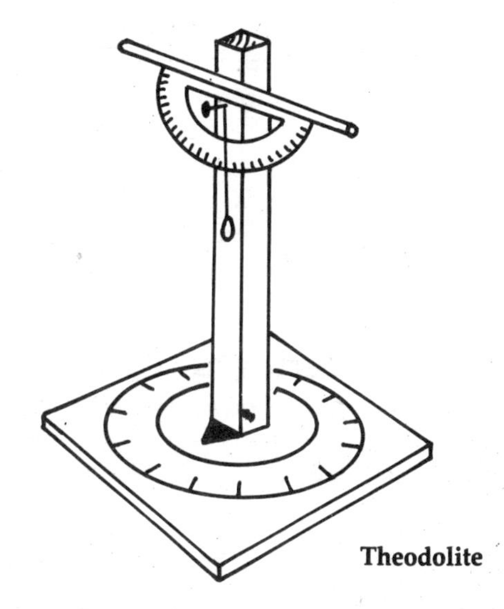

Theodolite

Position the full-circle protractor in the center of the 8 × 8-inch (20 × 20-cm) plywood, and glue it in place. Allow the glue to dry overnight, and carefully drill through the protractor's center and the plywood. Insert the longer screw through the bottom of the plywood until it pokes out the opposite side. Place a washer over the screw; then, attach the upright post to the screw, arrow side down, so that the arrow sits flat against the plywood. Make sure the upright post turns freely against the plywood base.

Next, glue the plastic straw to the straight edge of the half-circle protractor. The straw should stick out about 1 inch (2.5 cm) from opposite ends of the protractor.

Using a table edge or workbench, turn the theodolite on its side, and position the flat side of the protractor 3 inches (7.5 cm) from the top of the upright post. Attach the protractor in the center with the shorter screw and washer. Allow enough of the screw to protrude for tying on the nylon string. Move the protractor back and forth to make sure it pivots easily against the post. Turn the theodolite upright, and tie one end of the nylon string to the screw protruding from the protractor. Attach the metal fishing weight to the other end of the cord, making sure it hangs freely.

Finally, cover the front of the flashlight with red cellophane, attaching it with a few loops of a rubber band. This will allow you to read measurements on your theodolite while your eyes are accustomed to the dark.

What Measures Up

Now make some measurements. Place the theodolite on a level surface, and orient it so that the 0-degree mark on the base protractor lines up with *true north*. Do this by sighting the North Star through the straw. Be careful not to move the base from this position during other sightings.

Rotate the upright post clockwise, and adjust the straw up or down for any object that suits you. The altitude of the object is determined by reading the number of degrees on the half-circle protractor directly behind the nylon cord. The azimuth is the number of degrees indicated by the arrow on the full-circle protractor, that is, the distance you had to turn the upright post away from 0 degrees to sight the object.

Record the coordinates, date, and time of each sighting. Try sighting the same objects during several weeks or months, tracing the movement of various celestial bodies as seasons change.

Refracting Telescope

You Will Need

For Telescope

- 1¼-inch (3.25-cm) diameter PVC black plastic piping, 24 inches (61 cm)
- 1½-inch (3.25-cm) diameter PVC black plastic piping, 30 inches (76.2 cm)
- 2-inch (5-cm) diameter PVC black plastic piping, 3 inches (7.5 cm)
- 1 objective and 1 eyepiece lens
- Plywood piece 3 × 24 × ½ inches (7.5 × 60 × 1.25 cm) for support
- 3 thin brass strips ¾ × 5 inches (1.9 × 12.5 cm)

For Folding Tripod

- Plywood piece 6 × 6 × 1 inches (15 × 15 × 2.5 cm) for platform
- Hardwood block 2 × 4 × 1 inches (5 × 10 × 2.5 cm) for mounting block
- 3 pieces of 1 × 1 (2.5 × 2.5 cm) wood 43 inches (107.5 cm)
- 3 pieces of 1 × 1 (2.5 × 2.5 cm) wood 26 inches (65 cm)
- 3 small cabinet hinges
- 6 pieces of brass strapping ¹⁄₁₆ × 1 × 5 inches (.15 × 2.5 × 12.5 cm)
- Metal U bracket 1 × 2 × 1¾ (2.5 × 5 × 4.4 cm)
- 3 wood screws 1½ inches (3.25 cm)
- 5 wing-nut screws, 3 inches (7.5 cm)
- Three 1¼-inch machine bolts with washers and wing nuts
- 8 round-head wood screws ¾ inch
- 2 wood screws ½ inch (1.25 cm)

Additional Materials

- Thick felt 3 × 6 inches (7.5 × 15 cm)
- Electrical tape
- Carpenter's glue
- Protractor

Refractor & Reflector Telescopes

We cannot be sure who invented the first telescope. Some sources suggest the Dutch spectacle grinder Hans Lippershey, who constructed a simple magnifying device in 1608. But the first instrument of high quality was constructed by the Italian physicist Galileo Galilei (1564–1642).

Today, telescopes are classified into two types—refractor and reflector. The **refracting telescope** uses lenses (instead of mirrors) to refract (bend) light waves. Refracting telescopes are **Galilean** or **astronomical**, depending on the lens used for the eyepiece and whether the image is inverted.

Galileo's telescope interests us because he used a *double concave lens* for the eyepiece.

But most modern refracting telescopes use *double convex*, or magnifying-glass style, lenses. These are the astronomical telescopes, one of which we'll construct.

Refracting Telescope

Tubing & Lenses

This telescope uses nested pipes of plastic PVC (polyvinyl chloride) for the body, secured to a narrow support strip with brass strips.

Since the lenses make up the most important part of your telescope, choose them with care. You can find lenses in second-hand shops, optician-supply houses, or science supply stores. Lenses should roughly match the inner diameter of your tubes—1½ inches (38 mm) for the telescope's objective (wide side) and 1¼ inches (31 mm) for the eyepiece.

Lens Facts

Think of lenses as two prisms placed back-to-back. When parallel rays of light enter the prisms from one side, they emerge from the other side bent sharply towards each other. If you smooth this double prism arrangement into a double convex lens, the rays of light still emerge bent towards each other, but at a less severe angle. Where the rays of light converge is the **focal length** of the lens.

Determining focal length for the lenses you choose is important for getting the best image from your telescope. To do this, shine a bright light through the lens, and move the lens until the light focuses into a bright dot. Measure the distance from lens to dot to obtain the focal length. Your **objective lens** should have a long focal length (at least 39 inches or 1 m), and your **eyepiece** lens a short focal length (1 inch or 2 to 3 cm).

Unfortunately, inexpensive lenses usually have different focal lengths for each color of the spectrum. This defect creates a small degree of color distortion around the edges of viewed objects. *Achromatic lenses* correct this but cost considerably more. Replace less expensive lenses with costlier ones later, if you wish. Use simple lenses until you perfect your telescope-building technique. Since lenses also require periodic cleaning, attach them in a way they can be easily removed.

Constructing the Telescope

Cut off a 2-inch (5-cm) piece from the ends of both pipes. Make a straight, clean cut through the pipe so that you can place the smaller piece against the larger one without rotating it for an exact fit.

Attach your lenses to the unsawed end of each smaller piece by simply dropping them into position. Each lens should fit snugly with about 1 cm (⅖ inch) of pipe collaring the lens. If the lens is too large, rest it on top, and secure it around the edge with electrical tape. If the lens is too small, wrap tape around the glass edge to act as a spacer until you get a snug fit inside the pipe. Do *not* glue the lenses in place because you'll need

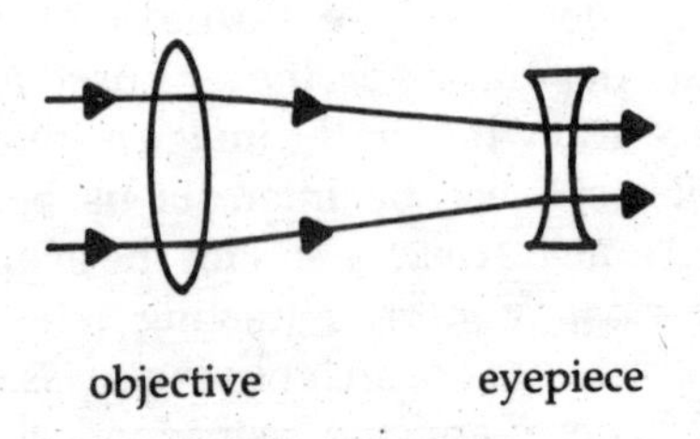

Galilean Telescope

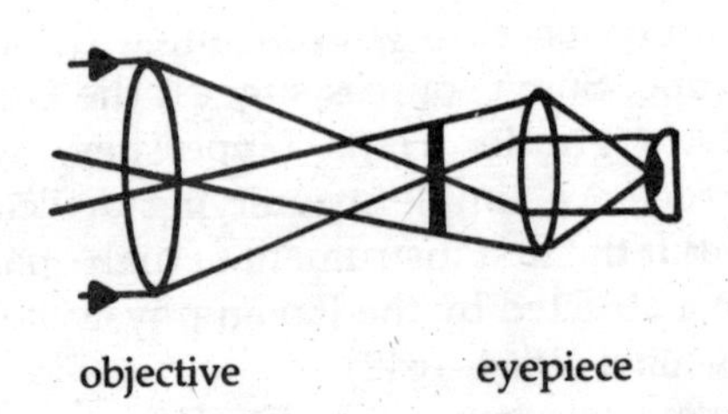

Astronomical Telescope

to test and adjust lens positions to ensure a clear image.

Tape the smaller pieces back onto the longer pipes, wrapping the tape several times for a tight fit. These removable pieces will come in handy when you want to clean your lenses.

Attach a hood for the telescope's objective to keep out extraneous light. Line the 3-inch (7.5-cm) pipe's inside with felt so that it fits snugly over the end of the wider tube and projects about 1½ inches (3.75 cm) from the end.

To assemble the telescope, insert the narrow pipe into the wide one, testing to make sure it slides easily. There should be just enough friction to prevent it from slipping out when the telescope is steeply inclined. If the fit seems too snug, use fine-gauge sandpaper to reduce the outside diameter slightly. If it slips or does not stay still when inclined, glue a strip of felt around the outside circumference to act as a spacer.

Folding Tripod

The folding tripod has three components—adjustable legs, circular platform, and a revolving mounting block with a metal bracket.

Adjustable Legs

Construct each leg with the long and short piece of 1 × 1 wood. Saw a 2-inch (5-cm) piece (wing) from the tops of the longer posts. Use a mitre box or table vise for the next step.

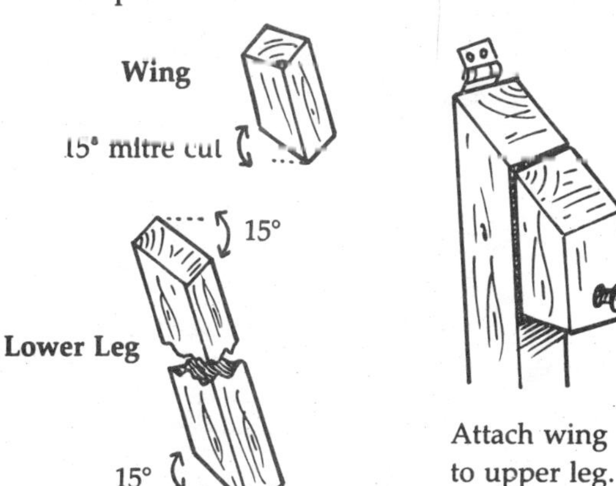

Attach wing to upper leg.

Make a 15-degree mitre cut at one end of each wing and each post. Turn each post upside down and make another 15-degree mitre cut, parallel to the first. Now, turn the wings around so that the mitre cuts of wings and posts line up, slanting down. Connect the wings to the posts with carpenter's glue and 1½-inch (3.25-cm) screws.

Finally, on the inside top edge of each completed leg (opposite the wing), locate the ½-inch (1.25-cm) center point, and mark it with pencil.

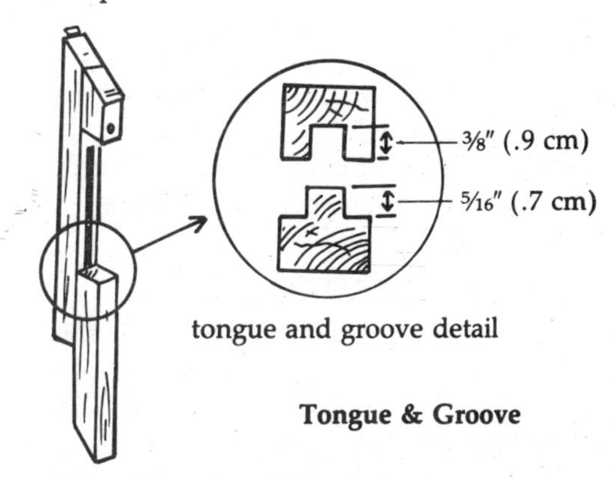

Tongue & Groove

Following the dimensions shown in the diagram, cut a groove in the longer piece (on the wing side) and a tongue in the shorter piece so that the posts fit together and can slide against each other. Although it's possible to cut a groove and tongue with certain types of hand planers, you're better off finding a lumberyard with a good dado saw. The groove in the longer piece should stop 1 inch (2.5 cm) from the bottom of the wing.

Clamp the pieces of each leg together with brass strapping bent as in the diagram.

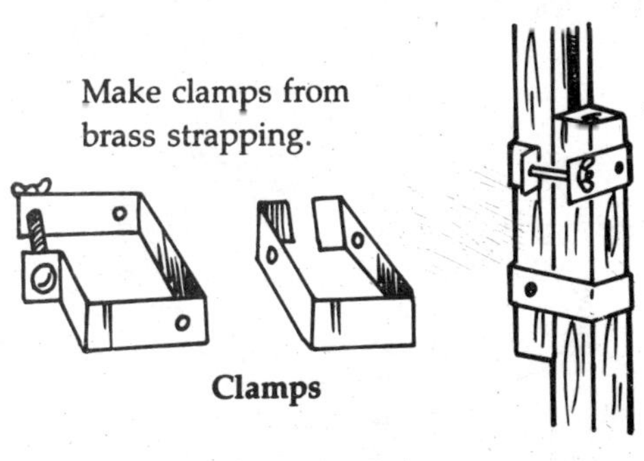

Clamps

Attach clamps to leg.

Notice the screw with wing nut attached to the larger clamp. Notice, too, drilled holes where screws will attach clamp to post. Secure the strapping in a table vise before bending it with pliers. You'll need six clamps for three legs—two clamps per leg.

Place the short post inside the long one. At 1 inch (2.5 cm) from the bottom of the long post, attach the smaller clamp, screwing it into the sides of the post as indicated. Six inches (15 cm) above this clamp, attach the larger clamp the same way, but make sure the wing nut faces out. Directly above the larger clamp, put a small round-head screw to prevent the two parts of the leg from sliding apart. The tripod legs can be easily adjusted by loosening and tightening the wing nuts on the larger clamps.

Circular Platform

For the circular platform, divide the plywood square in half with a horizontal pencil line. Follow it with a vertical pencil line. The intersection of the two lines represents the center of the square and will serve as the circle's center as well. Cut out a circle, 6 inches (15 cm) in diameter, from the plywood square, or have a lumberyard do it.

Draw a perfectly centered equilateral triangle as a guide for attaching the tripod legs. Here's how: Position the protractor vertically over the platform, centering it on the intersecting lines. The zero-degree line should overlap the vertical pencil line. Measure along the protractor to 120 degrees and mark it. Remove the protractor and draw a line from that mark to the center. Reposition the protractor so that zero degrees lies over the line you've just drawn. Measure another 120 degrees, mark it, and draw a second line connecting the mark to the center. Reposition the protractor over the second line, measure 120 degrees again, and draw a third line.

Now measure 1⅛ inches (2.8 cm) from the center point along all three lines and mark it. Draw the sides of the triangle by connecting these marks. You now have a centered equilateral triangle with 2-inch (5-cm) sides. Drill a hole in the center, wide enough for the wing-nut screw.

For the next step, place the platform on the floor. Locate the middle of each triangle side (1 inch or 2.5 cm from the corners), and make a mark. Align this mark with the mark on the inside edge of the first tripod leg. While holding the leg securely in place, position a hinge over the aligned marks and screw it in firmly, joining the leg to the platform. Repeat the procedure for the remaining two tripod legs.

Attach legs to platform.

tongue and groove

Tripod Construction

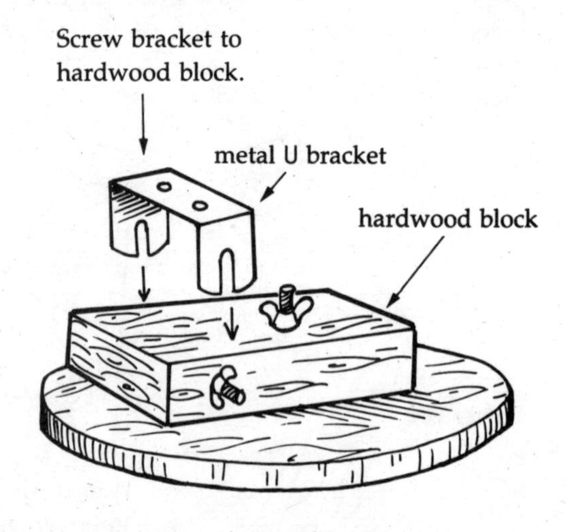

Tripod Platform (top)

Mounting Block

The 2 × 4 × 2-inch (5 × 10 × 5-cm) mounting block requires two holes—one drilled vertically, centered on top and 1½ inches (3.25 cm) from the narrow edge, and one drilled horizontally, centered along the side, 1 inch (2.5 cm) from the opposite narrow edge. Insert the first wing-nut screw vertically, up through the platform and the block, so that the wing nut faces up. Tighten the wing nut only so much—you want the block to rotate freely on the platform. Insert the second wing-nut screw horizontally through the block, attaching the bracket to it before tightening. By loosening and tightening these wing nuts, you will be able to fix your telescope in position.

Setting Up the Telescope

To set up the telescope, attach, with the metal bracket, the 3 × 24 × ½-inch (7.5 × 60 × 1.25-cm) plywood support piece to the mounting block. Then fasten the telescope to the long support with brass strips.

Turn the support on its side, and attach the three brass strips with the ¾-inch roundhead wood screws, spacing them evenly. On the opposite side, make parallel screw holes for the strips' free ends.

Lay the support piece flat, and measure 6 inches (15 cm) along its length, marking the point with a perpendicular line. Remove the metal bracket from the mounting block and place it, flat side down, against this line. Attach the bracket firmly to the support piece with two ½-inch screws.

Turn the support piece on its side again with the unattached ends of the brass strips facing up. Center the wider tube of the telescope against the support, and secure the tube by screwing down the free ends of the brass strips. Check to make sure the telescope fits snugly.

Testing the Telescope

Your finished telescope requires a final test to determine if the lenses need to be adjusted. Blurred images and extreme color distortions result when lenses are not perfectly parallel to each other. Point your telescope to a small, easily identified object and try focusing. You should obtain a sharp image with minimum distortion around the edges. Unsatisfactory results mean that you must pivot one or both lenses inside their pipes—testing again for clarity. If the results still remain unsatisfactory, remove and reposition the lenses.

Your telescope will provide many enjoyable hours of observation.

Sunspot Tracing

You Will Need

- Refracting telescope
- Card table
- Drawing pad with stiff cardboard backing
- Tracing paper
- Pencil compass
- Large black or dark-colored poster board
- Wide rubber band
- Duct tape

Sunspots

The sun's seemingly featureless face actually seethes with activity. Dramatic solar action can be seen in **sunspots**—those dark, cooler areas that seem to have seasons play havoc with earth's radio signals.

Sunspots and other forms of solar activity rise and fall in an 11-year **solar cycle.** The year 1986 marked the bottom of that cycle when sunspots were small and few; 1990 marked the peak when the sunspots were large and plentiful. But on any given day of the year, provided the sky is clear and the view is good, sunspots should be clearly discernable with even a low-power telescope.

Of course, you should *never look directly at the sun,* even through a telescope. Just a few seconds of the sun's intense light can damage your eyes. Various Mylar and glass filters exist, but they alter the color of the sun and have limited applications. For this project, we will collect and record sunspot data using the projection method.

Solar Observatory

The centerpiece of the setup will be the refracting telescope and tripod. Retract the lower legs of the tripod to make it as short as possible, about 3½ feet (105 cm) high. Place the card table near the telescope so that the tube's eyepiece hangs about 6 inches (15 cm) over the table's edge.

Place a stack of books about 5 inches (12.5 cm) from the eyepiece, and lean the drawing pad against the books so that the telescope's shadow appears. To keep the pad from slipping, secure it against the card table with a small, easily removable piece of

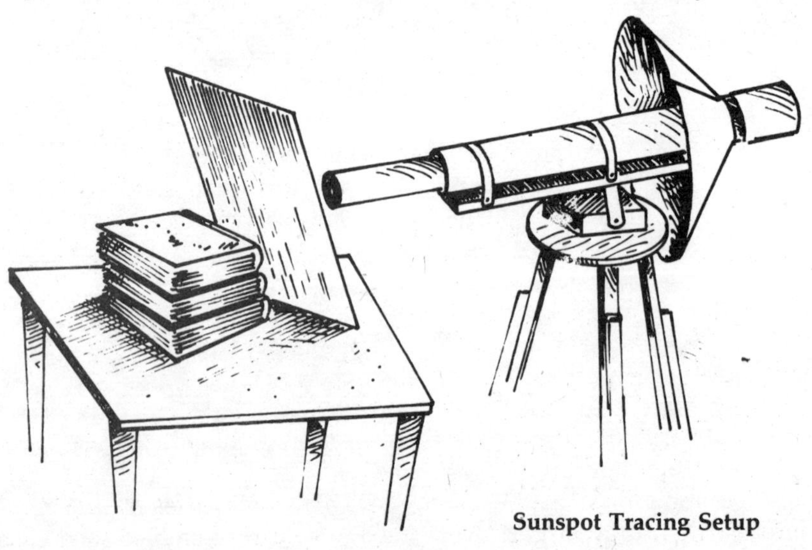

Sunspot Tracing Setup

duct tape. Make sure the pad leans rigidly against the books. If not, use cardboard reinforcement or a thin piece of plywood.

Choose a time of day when the sun is bright but not too high. Mid-morning and late afternoon are both good times for making observations. Maneuver the telescope while watching the shadow on the pad. When the shadow shortens and nearly disappears, your telescope is pointed directly at the sun and needs only fine adjustments before intense light pours through the eyepiece, producing an image on the pad.

But first, construct and attach a shade to the telescope tube. The shade filters out extraneous sunlight and improves the clarity of the image. Here's how: Cut an 18-inch (45-cm) diameter circle from the poster board (it need not be perfectly round). With the compass, trace a smaller 2-inch (5-cm) circle in the center. Draw a line from the edge of the poster board to the circle; then, cut along this line until you reach the circle and cut it out.

Carefully, to avoid shaking the telescope's position, insert the shade along the wide end of the telescope and support piece, pushing the poster board as far back as possible before you hit the tripod platform. Make sure the cut points down. Tape the cut ends together with a slight overlap so that the shade assumes a slightly conical shape around the tube. Finally, place duct tape around any gaps between the 3-inch (7.5-cm) circle and tube.

Spotting Sunspots

Tap the eyepiece side gently from side to side or up and down until you see a bright, round disk on the pad, from 4 to 6 inches (10 to 15 cm) in diameter. Move the pad closer or farther away, as necessary, adjusting the angle of inclination, to bring the disk's edges into sharp focus. When satisfied with the image's clarity, measure the distance from the eyepiece to the pad with a ruler, and write it down.

You may see sunspots, but ignore them for now. Trace the sun's disk on the pad using the compass, then draw a grid over the disk to plot the movement of sunspots later on.

Getting a good trace may take several tries. You'll also need to nudge the telescope as the image begins to drift from left to right, following the sun's movement across the sky. First, measure the disk's diameter by holding the ruler to the paper. Divide the diameter in half to get the radius, marking the circle's center. Tear the paper from the pad, out from under the eyepiece, and place it on a flat surface. Then, set your compass to the correct radius, place the needle end on the center mark, and draw a circle.

Using the ruler, cover the circle with grid lines ¼ inches (.62 cm) apart, making the lines as dark as possible. Align the top edge of a sheet of tracing paper to the top edge of the drawing paper. Draw a circle of the same dimensions on the tracing paper. Continue drawing circles on tracing paper for as many days as you want to observe the movement of sunspots—two weeks will suffice.

With the wide rubber band, reattach the grid to the pad under the eyepiece, and place the first piece of tracing paper over it, securing it with large paper clips, so that the circles overlap exactly. You should see the grid through the tracing paper. Mark the position of sunspots on the tracing paper with a soft pencil. Try to render as accurately as possible their sizes, shapes, contrasting shades, and graduated markings. Notice the **umbra** (dark center spot) and **penumbra** (lighter streaked area). When satisfied with your drawing, remove the tracing paper from the pad and write the observation date and time.

Repeat your observations at the same time on subsequent days, drawing sunspots on tracing paper placed against the grid. Individual spots and groups of spots will change in appearance from day to day. Familiar spots will also drift across the sun's surface as it rotates. Do you see any pattern to the appearance, shape, and disappearance of the sunspots? Do the spots seem to occur more frequently at a certain solar latitude?

Display your series of sunspot drawings next to your reassembled solar observatory.

Micrometeor Collecting

You Will Need

- 2 shallow glass pie dishes, including a heat-resistant Pyrex dish
- 1 quart of distilled water
- Strong magnet
- Small cellophane bag
- Hot plate
- Microscope
- 2 microscope slides and covers
- Specimen-mounting glue
- Disposable eyedropper
- Sewing needle

Frosty Meteors & Meteorites

Studying meteorites yields important information about the universe. But finding good specimens, especially in populated areas, can be a challenge. That's why astronomers regularly travel to places like the South Pole, where Antarctic winds of up to 200 miles (322 km) per hour blow meteorites across sheets of ice. The meteorites collect at the base of mountains, where they're easily recognized.

Meteors do not fall hot. Although they make a fiery appearance as they enter the atmosphere, only their surface is heated. Most meteors spend billions of years in space before falling to the ground; so, their interior temperature is close to absolute zero. In 1917 a farmer in Colby, Wisconsin, tried to pick one up and was rushed to a doctor with severe frostbite!

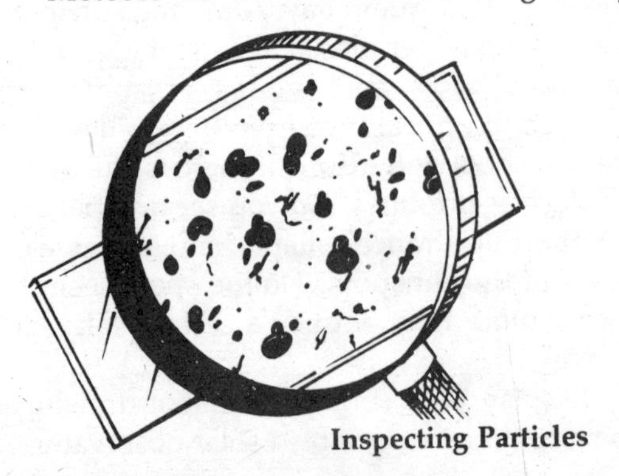

Inspecting Particles

Micrometeors

Though it's unlikely you'll find a meteorite in a suburb or city, you can easily collect **micrometeors**. Found everywhere, micrometeors are too small and too light to burn up as they enter the earth's atmosphere, so they remain floating in the air. Some fall to the ground only when attached to tiny water droplets or dust particles.

Like larger meteors, however, micrometeors can be either *metallic* or *rocky* and are mostly made up of debris left by comets and stellar explosions. That's why the best time to go micrometeor hunting is shortly after a meteor shower.

Check the chart Major Meteor Showers for appropriate times of year for major showers and for the estimated meteors falling per hour during each shower.

Magnet Sweep

The peak times for the best showers, **Geminids** and **Perseids**, occur December 13 and August 11, respectively. So plan to carry out the first phase of your project—collecting rainwater—within a 2-week span of these peaks. If you do not live in an area where it rains regularly, place a dish of distilled water outside during the same 2-week period.

Making a Collection

Make sure the glass dish is clean, washing it thoroughly with soap and water to remove dust particles before you set it outside. After enough rainwater collects or the filled dish has been left outside for a few days, cover a magnet with a small cellophane bag and place it in the dish. Slowly sweep the covered magnet along the bottom and sides of the dish. Because micro-meteors are rich in iron, they will be attracted to the magnet.

Carefully remove the magnet from the first dish, and place it in a second dish, filled with distilled water. Remove the bag from the magnet and gently swirl it around to allow micro-meteors to fall to the bottom of the dish.

Metallic Particles

Evaporate the distilled water so that micrometeors can be collected and mounted for observation. Place the Pyrex dish on the hot plate, and allow water to boil away completely. Next, magnetize the sewing needle by rubbing it against the magnet in one direction for about a minute. Drag the magnetized needle against the bottom and sides of the Pyrex dish.

Carefully tap the needle onto a clean microscope slide and mount the slide on the microscope for viewing. If you have a sufficient number of micrometeors on the slide, place a drop of mounting glue over the area and lower the slide cover over the glue. Label the slide "metallic particles" and allow it to dry.

Nonmetallic Particles

Next, separate the nonmetallic particles. To do this, evaporate the water in the original dish. Using the needle again (its magnetization is unimportant now), scrape the bottom and sides of the dish and then tap the needle onto a clean microscope slide. Mount the slide and check to see if you have enough particles. If so, glue a cover on the slide and label it "nonmetallic particles."

Carefully compare the two slides, noting the differences between the particles. Draw a few particle types—rounded particles and jagged ones; large particles and small. Jagged, similarly shaped nonmetallic particles probably come from the earth, but rounded metallic particles are probably extraterrestrial in origin.

Position your microscope prominently, and invite viewers to see your fine collection of micrometeors.

MAJOR METEOR SHOWERS

Shower	Date of Maximum Number of Meteors	Hourly Rate for a Single Observer
Quadrantids	January 3	40
Lyrids	April 21	15
Aquarids	May 4	20
Aquarids	July 28	20
Perseids	August 11	50
Orionids	October 21	25
Taurids	November 4	15
Leonids	November 16	15
Geminids	December 13	50
Ursids	December 22	15

Detecting Cosmic Rays

You Will Need

- Straight-side glass jar with metal lid
- Heavy felt
- Black velvet
- Rubber cement
- Rubbing (isopropyl) alcohol
- Pie tin
- Small magnet
- Ice tongs
- Flashlight or slide projector beam

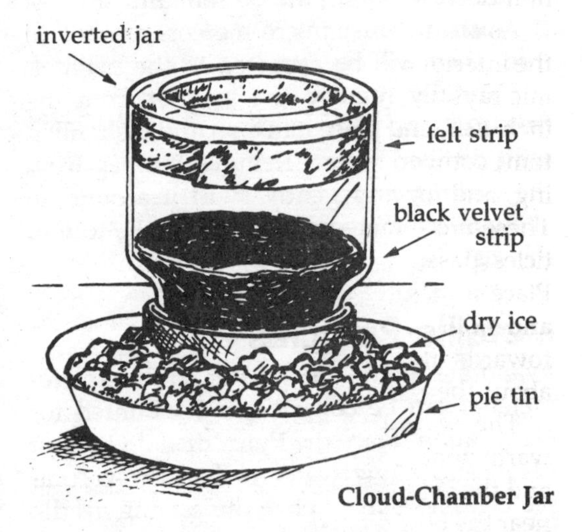

Cloud-Chamber Jar

Cosmic Rays

At this moment, tiny particles left over from the explosion of stars are hurtling through space at nearly the speed of light. Trillions of these particles—called **cosmic rays**—pass through the earth's atmosphere every few minutes, and though too small to see, you can detect their presence with the aid of a **Geiger counter** or **cloud chamber**. We'll construct a cloud chamber.

Cloud Chamber in a Jar

Select the jar for your chamber carefully. A short, widemouthed jar with clear sides and bottom works well, particularly if it has a rubber washer or cardboard filler under the lid.

With a little rubber cement, stick the strip of heavy felt around the inside wall at the bottom of the jar. Cut a strip of black velvet 1 inch (2.5 cm) wide and glue it to the top of the jar's inside wall. Cut a circular piece of velvet just large enough to fit inside the metal lid, over the rubber or cardboard, and glue it in place. Pour enough rubbing alcohol into the jar to saturate the felt thoroughly, and with leftover alcohol just cover the bottom of the jar. Screw the lid on the jar and tighten it.

Dry-Ice Stand

Be especially cautious, or ask an adult to help with the dry ice, a substance that's dangerously cold to touch. First, spread a towel on a flat surface and place the pie tin on top of it. Now, using ice tongs, carefully put the cubes of dry ice into the tin. If the ice comes in a block, wear rubber gloves to wrap a heavy cloth around it, then break up the block with a hammer. Check to see if the ice makes a level surface over which you can rest the jar.

Place the jar on the dry ice with the metal lid down. Position the flashlight on a book or two, attaching it with masking tape if necessary, so that the beam goes right through the jar's center. Cover the bottom of the jar (what's now the top of the cloud chamber) with the palm of your hand to warm the interior slightly. Watch carefully.

Ghostly Trails

When the alcohol vapor condenses, warmed by your hand, you'll see a continuous rain of fine droplets 1 inch (2.5 cm) below the top of the chamber. Be patient, because this may take as long as 5 minutes. After another 5 minutes or so, the amount of this rain decreases.

As conditions become right, that is, when the interior temperature is just right and cosmic rays are present, you will see about an inch (2.5 cm) above the lid (now the bottom) cobweb-like threads suddenly appearing and disappearing at various angles. These are vapor trails made by cosmic particles passing through the cloud chamber. Place a magnet against the side of the jar, and notice how the trails appear deflected towards it. Does this suggest something about the particles' composition?

The warmth from your hand creates a warm zone towards the top of the chamber, and the cold of the dry ice makes a cool zone near the bottom of the chamber. Somewhere between these temperature extremes, usually about 1 to 2 inches (2.5 to 5 cm) from the bottom, the air becomes saturated with alcohol vapor. That's where to look for particle trails.

Photos

You could photograph these particle trails by setting up a camera at the viewing angle and taking a time-exposure photograph. Display the photograph beside your project to give your viewers a time-lapse overview of how cosmic rays behave inside a cloud chamber.

Dry Ice Caution

Remember, dry ice will cool the pie tin, making it hazardous to touch. When you've finished your display, use pot holders to lift the tin from the ice. Then carefully remove the ice with tongs and dispose of it in a safe place. Finally, allow the tin time to return to room temperature—at least an hour—before handling it.

Gravity & Curved Space

You Will Need

- 14-inch (35-cm) embroidery hoop
- Plastic wrap
- 1-inch (2.5-cm) large marble or steel ball
- 10 small ball bearings or BB's

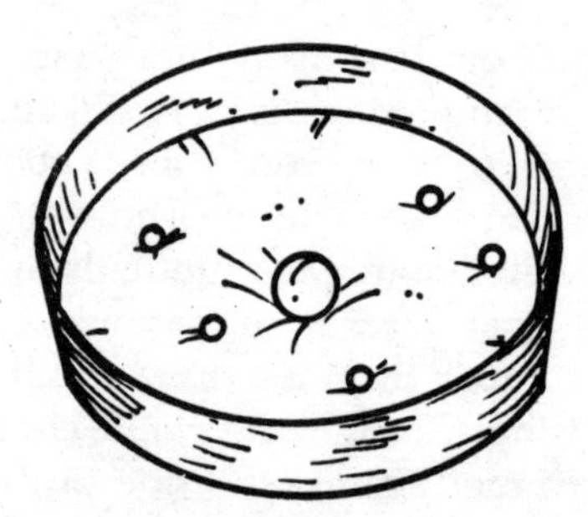

Cellophane Drum with Marble

A Matter of Gravity

According to **Einstein's theory of relativity,** gravity is not a kind of magnetism as Newton thought, but a *distortion in the curvature of space* due to the presence of massive concentrations of matter. This distortion affects how lesser concentrations of matter behave around greater ones.

BB Test

Stretch the plastic wrap across the embroidery hoop until you have a tight, drum-like surface. This represents an empty and vacuous universe. Scientists believe that space between galaxies is very much like this.

While holding the hoop at its edges, have someone roll a BB across the surface of the stretched membrane. Notice how the weight of the BB causes a slight local distortion in the drum surface. But the BB rolls in a straight line because no other mass is present to create another distortion.

Now place the large marble in the center of the membrane to represent an object with a huge mass. The object's increased weight noticeably distorts the membrane surrounding it.

Roll one BB from the hoop's edge towards the marble. The BB either goes into orbit around the marble or deflects towards it as it passes. Try rolling BBs at different speeds and in different directions towards the large marble. With a little practice you should be able to create perfectly formed, though short-lived, elliptical orbits around the marble.

After the **big bang**—the tremendous explosion that created the universe—galaxies, planets, and stars began to form as diffuse particles of matter came together, attracted by mutual gravitation.

To understand this process, try this demonstration. Remove the large marble, and place 10 BBs at random points on the drum. Gently shake the hoop back and forth so that the BBs move without hitting the drum's side. Because of this random motion and the local distortions of each BB, they will soon be attracted together to form one large mass.

Ecology

How Clean Is the Air?
Airborne Acid Corrosion
Melting Marble
Detergents & Pond Life
Shrinking Water Table
Water Retention of Soil
Oily Animals
Testing Biodegradability
Overpopulation Studies
Light Pollution Diorama

How Clean Is the Air?

You Will Need

- 14 white index cards
- Jar of petroleum jelly
- Magnifying glass

Modern Pollution

Each day, millions of different kinds of **particles** find their way into the air we breathe. Some, such as soil or salt particles, are harmless. Others, like plant pollens, may be only irritating. But particles from industrial waste and automobile exhaust pose a serious health threat and have made that breath of truly fresh air a rarer experience for most people.

Measuring Particles in Air

You can estimate the **pollutant level** in your area and determine both the kind of particles and the pattern of their concentration.

Divide the 14 index cards into two groups of seven. Label the first set of cards 1 to 7 to represent each day of the week. Label the second set of cards 1A to 7A. Using a ruler, draw a 1-inch (2.5-cm) square in the center of each card with a felt-tip marker. With your finger, or a cotton swab, smear a small patch of petroleum jelly into the square, making sure the jelly is evenly spread and not too thick.

Tape card 1 to the *inside* of a window, with the petroleum jelly square facing you. Tape card 1A to the *outside* of the window, with the petroleum jelly square facing away from you. Note the time you attach the cards to the window. After 24 hours, remove the cards and replace them with card 2 on the inside of the window and card 2A on the outside.

Now compare card 1 and 1A by examining each closely with a magnifying glass. Which petroleum jelly square appears "dirtier"? Note the number, size, and shape of the largest particles on card 1 and record your observations. Now note the difference in both the shapes and quantity of particles on card 1A. Indoor pollution consists of dusts, fibers from clothes or carpets, animal fur, and fine particles of smoke. Outdoor pollution consists of larger ash and carbon particles, soil, fungi, plant pollens, and sometimes salt particles from ocean spray.

After 24 hours, remove and replace cards 2 and 2A with cards 3 and 3A, and so on, until you have two complete sets of cards for the week. Each day, compare the sets of cards not only to each other, but to the previous set, recording all observations. Also note conditions both inside and outside while your cards collect particles. For instance, does a windy day affect the number and size of particles? Or a rainy day? If the room is vacuumed and dusted, does this affect the number and size of particles? If conditions remain the same throughout the week and your cards seem to collect more particles on a Thursday or Friday than they did on a Monday or Tuesday, does this suggest something about the pattern of industrial activity in your area? How do the particles collected over a weekend differ from the particles collected during the week?

Repeat your experiment in a less populated area and compare the results. Overall, how clean is the air in your neighborhood?

Airborne Acid Corrosion

You Will Need

- Old nylon stocking
- Two 8½ × 11-inch (21.25 × 27.5-cm) pieces of stiff cardboard or poster board
- 8½ × 11-inch (21.25 × 27.5-cm) paper
- Craft knife
- Magnifying glass
- Cellophane tape

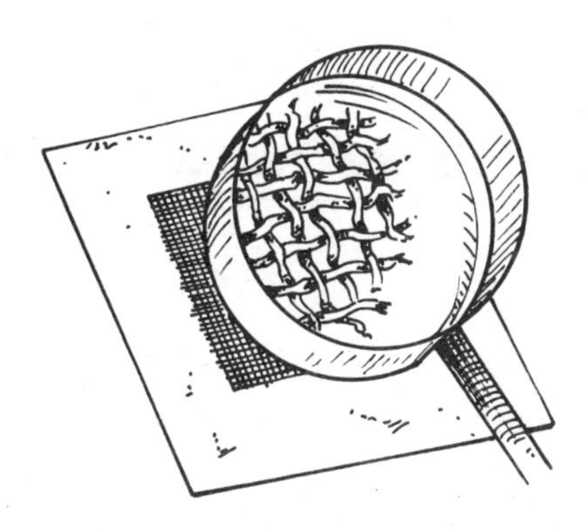

Corrosive Times

In many cities around the world, statues, monuments, and magnificent old buildings display the destructive effects of **pollution.** Some have been damaged more in the last 50 years than in all their previous history. The damage consists of corrosion from **acid fumes,** produced by manufacturing and automobile exhausts. Such fumes—dry or mixed with water to create **acid rain**—eat through stone, metal, and paint, and can weaken or destroy rubber, leather, and other fabrics. The effect on our lungs cannot be healthy, either.

Nylon Test

It might seem hard to believe the air we breathe contains corrosive properties, but this project demonstrates the seriousness of the problem.

Using a ruler and marker, draw a 4 × 6-inch (10 × 15-cm) window inside both pieces of cardboard and cut them out, using the craft knife. With scissors, cut two pieces of nylon stocking large enough to fit over the windows, taping both pieces securely in place.

Examine the nylon through the magnifying glass, noting the threads' regular weave and uniform thickness. Compare both pieces. Pinch the nylon and gently stretch it to test the material's strength and flexibility. Then, put one of the cardboard pieces aside, and cover it with a sheet of paper.

Mount the second piece of cardboard outdoors where the air can blow through it. Or prop it under a window against the windowsill if you have a steady breeze.

After about 2 weeks, remove the exposed cardboard, and examine it again with the magnifying glass. Uncover the second cardboard piece and compare it with the exposed piece. Do the nylon threads on the exposed piece look the same? Gently stretch the nylon of the exposed piece. Does it seem as strong or as flexible as the nylon of the covered piece?

At a glance you'll see damage from acid fumes in the threads of the exposed nylon. In some places, the threads seem to have melted away or turned black and brittle. In other places, they have disappeared completely, leaving holes in the fabric.

For your exhibition, mount both pieces of nylon with the magnifying glass handy. Record the location—where you live, conducted the experiment, and mounted the cardboard. Also record the length of time you exposed the nylon to air. You might want to create a local map of your area, highlighting industrial regions which may have contributed to the acid fumes.

Melting Marble

You Will Need

- Small marble tile
- Vinegar
- Teaspoon
- Calcium carbonate (powdered marble)
- Bromothymol blue solution
- Phenolphthalein base indicator
- Two 50-ml beakers
- 200-ml beaker or jar
- Erlenmeyer flask with one-hole rubber stopper
- Short piece of glass tubing
- 18-inch (45-cm) length of narrow rubber hose

Acid Rain & Marble

The destructive capabilities of **acid rain** have been well documented. But why is marble—the material of so many masterpieces in art and architecture—particularly vulnerable? This project provides the answer.

Bubbling Pool

Begin by placing a few drops of vinegar on the marble tile, enough to make a small pool. After about five minutes, note the tiny bubbles forming inside the pool. The acidic vinegar reacts with the marble to release a gas. Wipe the marble clean of vinegar and examine the surface. Does the marble appear duller in the area of the vinegar pool?

In powdered form, marble is called **calcium carbonate**. Chemical reactions become more visible when you use powdered marble rather than solid chunks of the stone. In the next step, we'll identify the gas released when the vinegar reacts with the marble tile.

Erlenmeyer Flask & Beaker

Naming the Gas

Prepare your test by pushing the short glass tube through the hole in the rubber stopper. Connect the rubber hose to one end of the tube, where it emerges from the wider side of the stopper. Place the stopper and tube assembly next to the Erlenmeyer flask and beaker.

Fill one of the beakers halfway with the bromothymol blue solution. Put 2 or 3 teaspoons (10 to 15 ml) of calcium carbonate in the Erlenmeyer flask and add 50 ml of vinegar. Immediately cap the flask with the rubber stopper, and place the end of the rubber tube into the beaker. Bromothymol blue is a **chemical indicator** which reacts to the presence of carbon dioxide by changing color. Does the bromothymol blue change color? We can now safely conclude that acid reacts with marble by producing carbon dioxide.

But what about acid rain? What causes the marble to melt away as though it were made of something far less sturdy? We'll find out why acidic substances are particularly destructive to marble in the next part of the project.

Opposites Destroy

Mix 1 teaspoon (5 ml) of calcium carbonate with 50 ml of water in a 200-ml beaker. Add five drops of phenolphthalein to 50 ml of water in a separate beaker. Add this solution to the calcium carbonate beaker. Like bromothymol blue, phenolphthalein indicates the presence of a chemical *base* by changing color. As you add the phenolphthalein to the calcium carbonate solution, notice the color change from clear to pink. From this, we can conclude that marble is a base. Adding a little vinegar to the pink solution turns it clear again. The base solution of dissolved marble neutralizes the acid of the vinegar.

Such a reaction quickly dissolves marble. Courageous steps need to be taken to preserve some of the ancient monuments of Western civilization from the recent ravages of acid rain.

Detergents & Pond Life

You Will Need

- Pond plants, such as elodea or algae
- Pond animals, such as daphnia (water fleas)
- Six 12-ounce (354-ml) mayonnaise or jam jars
- High-phosphate laundry detergent
- Measuring cup
- Eyedropper

Detergents & Pond Test

Old-Fashioned Soap

What would we do without soap to wash our clothes and dishes? But from the beginning, detergents—so effective at removing oil and grime from our lives—have taken a toll on the environment.

Old-fashioned (nonphosphate) soaps were slow to break down in water or soil and eventually created billowing clouds of suds around sewage plants, private wells, and waterways. Soon, people found that their tap water was soap-polluted as more and more detergents seeped into the underground water table. The solution was a new detergent containing chemical **phosphates**.

Phosphate Side Effects

Phosphates are quick to break down, but they have had an unanticipated side effect related to their fertilizing properties. **Algae,** exposed to phosphate-rich water, multiplied so rapidly that it used all available oxygen and suffocated other pond life. Soon, this process of **eutrophication** turned even the most thriving ponds into stagnant, foul-smelling messes. Now scientists are working on developing nonphosphate detergents while carefully monitoring the use of phosphates in our environment.

Pond Test

You can demonstrate the detrimental effect of phosphates on freshwater life. If you cannot find a stream or pond in your area, buy 6 small elodea plants and a jar of daphnia from an aquarium shop. Grow algae by placing a jar of aquarium water in the sun for a few days or until it begins to turn green.

Place the 6 jars side by side and fill each halfway full with the algae-green water. Drop an elodea plant into each jar and add a few daphnia with the eyedropper. Mix ½ cup (120 ml) of detergent with 1 pint (500 ml) of tap water.

With the eyedropper, place 5 drops of the detergent mixture in the first jar, 10 drops in the second, 20 drops in the third, 40 drops in the fourth, and 80 drops in the fifth. Use the sixth jar as a control to which you add *no* detergent.

Place the row of jars in a sunny place and observe them each day. Notice how the algae responds to the higher concentrations of detergent at the expense of the elodea and daphnia. In what jar does the concentration of algae make all other life impossible (eutrophication)? Compare the odors from each jar and correlate them with the amount of algae present. Finally, label each jar so that your viewers can see which contain the higher levels of phosphates.

Shrinking Water Table

You Will Need

- Medium rectangular aquarium tank
- 8-inch (20-cm) large-bore glass or plexiglass tubing 2-inch (5-cm) diameter
- 4-inch (10-cm) small-bore glass or plexiglass tubing 1-inch (2.5-cm) diameter
- Glass jar or beaker
- Glass T tube
- Two glass straws or pipettes
- Two 18-inch (45-cm) lengths of narrow rubber tubing
- Rubber bulb from ear syringe or poultry baster
- 1-pound bags of sand, gravel, and soil
- Craft knife
- Dishwashing detergent
- Sewing needle and short thread
- Small piece of balsa wood
- Small fisherman's weight

Nor Any Drop to Drink

Every living thing on earth depends on clean, healthy water to survive. The greatest source of that water—the **water table**—lies underground. Continued depletion and pollution of this precious natural resource foreshadows serious future consequences.

The water table can be damaged in many ways. Irresponsible dumping of bacterial and chemical waste products can cause toxic substances to accumulate and leach through the ground, eventually tainting the water we drink and bathe in. Overpumping for irrigation and forced drainage for land development deplete the store of underground water. This results in dried-up lakes, ponds, and streams, and destruction of valuable marine life.

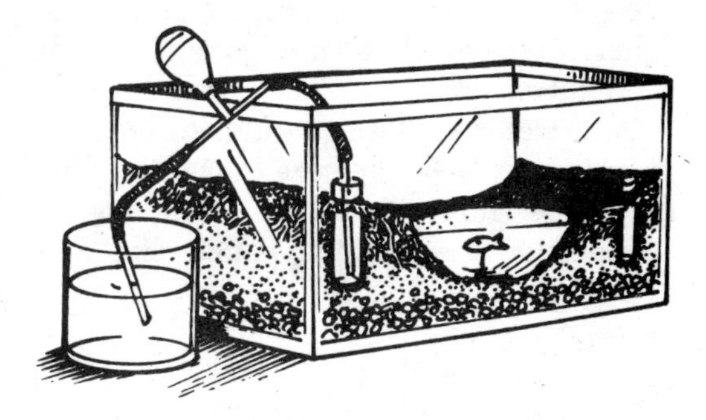

Water-Table Model

Model Water Table

You can dramatically demonstrate the effects of a compromised water table.

Cover the aquarium's bottom with a layer of gravel about 3 inches (7.5 cm) thick; then, gently slope the layer up towards the left side of the aquarium. Follow with a 3-inch (7.5-cm) layer of sand, also sloped. Finish with a 3-inch (7.5-cm) layer of soil. These layers represent the earth's composition above bedrock. Using the spray bottle, moisten the top layer so that you can mold the soil more easily.

With the teaspoon, dig a small pond in the middle of the aquarium, close against the front glass so that you can see inside the pond. Make the pond no more than 10 inches (25 cm) across. Next, dig a hole 2 inches (5 cm) wide and 6 inches (15 cm) deep in the right corner against the glass. Insert the large glass tube into the hole. Dig another hole 1 inch (2.5 cm) wide by 3 inches (7.5 cm) deep at the opposite corner against the glass. Insert the smaller tube there.

Carve a small fish from the balsa wood with a craft knife, and anchor it at the pond's bottom. If you're unfamiliar with carving, have an adult help you. When you've completed your fish, measure a thread about ½ inch (1.25 cm) shorter than the pond's depth. Thread the needle, knot the thread, and push the needle completely through the fish from the bottom. Attach the free end of thread to the weight and place the weighted fish in the middle of the pond. For realism, add small plants to your aquarium-terrarium.

Attach the 18-inch (45-cm) rubber tubing to opposite ends of the T tube, and attach the rubber bulb to the tube's stem. Stretch the ends of the tubes over the glass pipettes. Fill the beaker or jar to the top with water, and place one of the glass pipettes inside. Place the other pipette in the large tube or deep well of the aquarium. Clamp off the aquarium hose with a large paper clip near the T tube.

Add water to the aquarium to create a visible water table. Squeeze the bulb so that the rubber tube fills up to where you've clamped it. Remove the paper clip, and place it on the opposite side of the T tube. Squeeze the bulb so that the water runs into your deep well. Repeat the procedure, little by little, until the layers of gravel, sand, and soil become saturated and water begins filling the pond and shallow well (smaller glass tube). Your balsa-wood fish swims merrily.

Now reverse the operation, pulling water out of the aquarium and back into the jar. Watch the pond emptying until the fish settles on the bottom. Also notice how the sides of the pond appear to collapse as drying soil tumbles from the walls. What do you think would happen to the pond with repeated filling and drying?

Fish Bubbles

The next part demonstrates the effect even a small amount of polluting material has upon the water table. After you've drained the aquarium, place a few drops of dishwashing detergent in the shallow well. Gradually replenish the water table as before, but what do you notice about the water in the pond? Do you think the effects of the detergent will diminish as you continue to raise and lower the water level in the aquarium? How do you think plants will respond? The health of our water table requires serious thought.

Water Retention of Soil

You Will Need

- 4 soup cans with tops and bottoms removed *or* 4 pieces of PVC pipe 4¼ × 2¾ inches (10.6 × 6.8 cm)
- 4 medium bowls
- Framed window screen about 1 × 2 feet (30 × 60 cm)
- 4 wooden leg blocks 2 × 2 × 6 inches (5 × 5 × 15 cm)
- Measuring cup
- Potting soil
- Coarse sand
- Potter's clay or pond clay
- Marbles
- 2 small flowerpots
- Grass seed
- Carpenter's glue
- 50-ml graduated beaker
- Stopwatch

How Green Was My Garden

Soil quality has everything to do with the health of trees, grasses, flowers, fruits, and vegetables. The best soil, **topsoil**, contains *nitrogen* from decaying plant and animal matter and usually is the top, thin layer in any soil mixture. A layer of sand or clay exists beneath topsoil.

It doesn't take much topsoil to create and sustain a healthy growing environment for plants, as long as the layer remains concentrated. Many developers remove topsoil or degrade good soil by mixing rich upper topsoil with the sand and clay beneath it. Then property owners replace it with topsoil brought from other places for landscaping.

Topsoil contains important minerals for plant nutrition and retains just the right amount of water to nourish plants without damaging them. Water passes through sandy soil too quickly, leaving plant roots little time to absorb. Water passes through clay too slowly, which leads to rotting roots and a drowning plant.

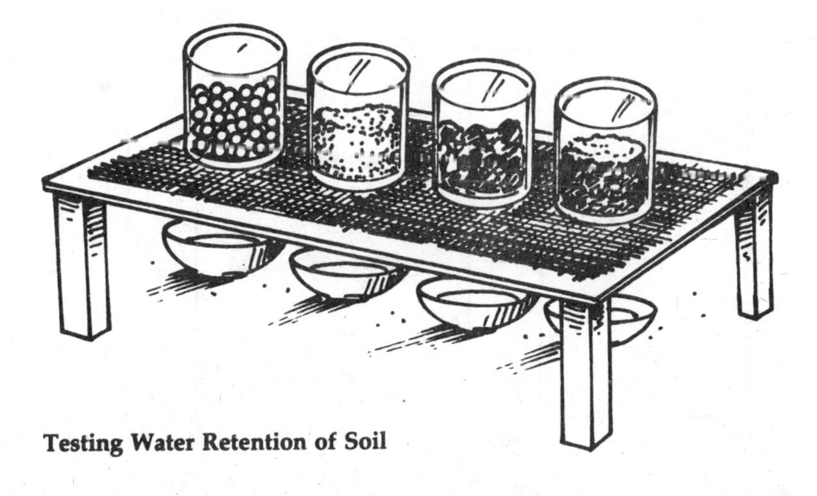

Testing Water Retention of Soil

Measuring the Right Stuff

We'll first compare water retention in topsoil, sand, and clay. Then, we'll compare degraded soil with layered topsoil to see which provides a healthier growing environment.

Place the framed screen on a flat surface, and attach leg blocks to each corner with carpenter's glue. Allow the glue to dry overnight before turning the screen over to rest on its legs.

Evenly space four cans on the screen. Place a bowl under the screen beneath each can. Fill the first can with 3 cups (720 ml) of marbles. Then place 3 cups (720 ml) each of potting soil, sand, and clay in the remaining cans. Label the cans with a felt-tip marker.

Our control, against which we'll measure results, is the can of marbles. We need to determine how much and how quickly a given quantity of water passes through a nonabsorbing medium.

Click the stopwatch as you quickly pour 50 ml of water from the beaker into the can containing the marbles. When the last drop of water passes through the can and into the bowl, click the stopwatch again and record the time. Then, pour the water from the bowl back into the graduated beaker, and record the water level. It should be nearly the same level as when you started.

Repeat the procedure for the cans of sand, clay, and potting soil. Record how long it takes water to pass through, and measure the amount of water recovered in the bowls by pouring the water back into the beaker.

Making Graphs

Display your data on graphs. Use one graph to indicate the time in seconds it took water to pass through the materials, and a second graph to indicate the millilitres of water recovered in the bowls.

From your graphs, you will easily see how sand retains water less successfully than potting soil or clay, and makes a poor soil environment for plants. Clay holds water too long. Potting soil seems just about right for plants' growing environment.

Repeat the procedure using samples of local soil instead of store-bought potting soil. How would you rate the soil in your area?

Spoiled Soil

In a bowl, mix a cup of sand, clay, and potting soil. Fill one of the two flowerpots with the mixture and add ½ teaspoon (2.5 ml) of grass seed. Cover the seeds lightly with more mixture.

At the bottom of the second flowerpot, place a layer of clay, followed by a layer of sand, then a layer of potting soil. Place ½ teaspoon (2.5 ml) of grass seeds in the potting soil, and cover them lightly.

Water both flowerpots and place them in a sunny area. Observe the grass stalks growing from each pot, comparing quantity, size, and thickness. Which pot appears to provide a healthier environment for grass? Repeat the experiment using other plants. Are the results constant?

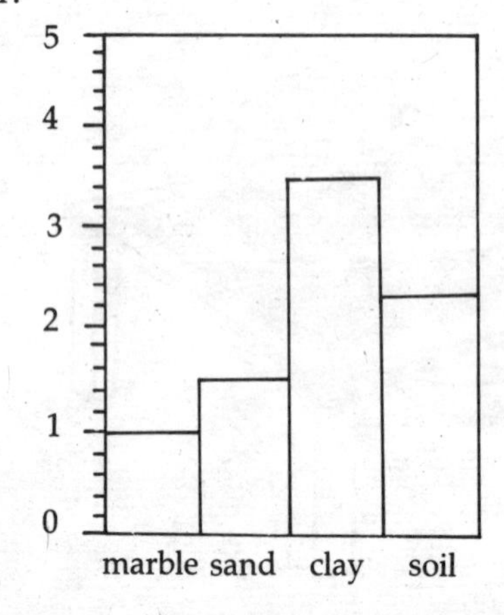

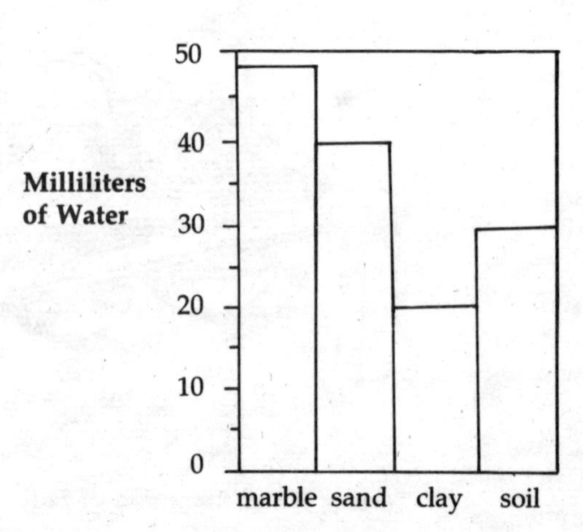

Oily Animals

You Will Need

- Light-colored rabbit fur, cut into three 2 × 2-inch (5 × 5-cm) squares
- Chamois cloth, cut into three 2 × 2-inch (5 × 5-cm) squares
- 3 light-colored feathers
- 6 bowls
- Graduated cylinder
- 40 ml (2⅔ tablespoons) of used motor oil
- 40 ml (2⅔ tablespoons) each of dishwashing liquid detergent, powdered dishwashing detergent, and powdered laundry detergent
- Sponge

Oil Pollution

Virtually every ocean on earth contains particles of **oil pollution**. Giant oil spills, whether from leaking supertankers or ruptured pipelines, create havoc in delicate saltwater ecosystems and kill thousands of animals. This project demonstrates the destructiveness of oil on animal life.

Bowl Test

Separate the six bowls into two sets of three. In one bowl, mix 40 ml (2⅔ tablespoons) of dishwashing liquid detergent with an equal amount of water. Repeat this procedure in the second and third bowls for the powdered dishwashing detergent and powdered laundry detergents, labelling the bowls with a felt-tip marker and masking tape.

Leave the second set of bowls unlabelled. Fill the first of the unlabelled bowls with clear water, the second with 40 ml (2⅔ tablespoons) of motor oil mixed with 10 ml (2 teaspoons) of water, and the third with clean water for rinsing.

Soak the three squares of fur in the bowl containing clear water. After about 1 minute, place them in the bowl containing oil and water. After about 3 minutes, remove them from the oil, and rinse them in the clean water. Repeat this procedure with the feathers and chamois cloth squares. Place all materials on paper towels.

Before you attempt to clean the fur, feathers, and chamois cloth, examine the pieces carefully, and note how the oil changes the texture of each material. With a dry paper towel, wipe off excess oil from each sample. Which material seems the most resistant to oil? Which seems the most damaged? Is it possible to remove oil from any of the samples without harming them?

Fur resists the oil more successfully than the feather or chamois samples. The bristles of fur catch and suspend the oil above the skin where it might cling and suffocate the animal. That's why animals with furry bodies, like otters, beavers, and seals, have a better chance of surviving an oil spill.

Less effective than the fur of the otter, the feathers of a tern, grebe, or sea gull hold the oil, which, in turn, destroys the feathers' delicate, waterproof fibers. A bird coated with oil cannot fly or float easily, and the oil easily seeps below to the skin where it may cause suffocation. An additional complication involves preening, since a bird covered with a foreign substance ingests the substance while cleaning itself.

The soft leather of the chamois cloth simulates the skin of large sea mammals, like the walrus or whale. Without any natural protection, these animals quickly become coated with any toxic spill, which clings to skin and is absorbed by their bodies.

Finding the Right Solution

Scientists, working to discover effective means for rescuing oil-covered animals, have invented solutions that effectively clean while doing little damage to the animal. The solutions were discovered by trial-and-error.

Place a fur, feather, and chamois cloth sample in the first labelled bowl containing dishwashing liquid detergent. Place additional samples in the second and third bowls containing powdered dishwashing detergent and laundry detergent, respectively. Allow the samples to soak for five minutes. Then, carefully sponge each sample in its own soapy solution, rinse in the bowl of clear water, and place them on paper towels to dry.

Compare the samples. Why do you suppose the concentrated powdered laundry detergent cleans off the oil more effectively than the liquid dishwashing detergent or powdered dishwashing detergent? Read the list of ingredients on the powdered laundry detergent box. Does it contain something lacking in the other two detergents? In fact, laundry detergents contain many chemicals that penetrate deeply into cloth fibers, and break up, or **emulsify,** oil, allowing it to be rinsed away. These chemicals have been isolated and adapted for cleansing oil-soaked animals.

Testing Biodegradability

You Will Need

- 3 widemouthed glass jars
- Bag of peat moss, mulch, or potting soil
- Plastic straw
- Glass marble
- Small nail
- Aluminum foil
- Cotton fabric
- Balsa or soft wood
- Large white poster board
- Magnifying glass

Hauling Out the Trash

Our modern industrial civilization throws away much more than it uses, and the problem of what to do with the trash challenges scientists. Mountains of garbage destroy the landscape's natural beauty and pose a health hazard to those living nearby. As many synthetic and petroleum-based products find their way into landfills, the problem becomes more urgent and complicated. Materials like plastic and glass resist decay and remain in the soil for hundreds of years. Other materials like wood, paper, and fabric, can be readily broken down by microorganisms and chemicals in the soil. These latter materials are **biodegradable**.

Many people have begun to recognize the importance of keeping nonbiodegradable materials out of landfills. One answer is recycling, and many large cities recycle products like glass and plastics. This project will test biodegradable properties of various materials.

How Biodegradable Is It?

Fill the 3 jars halfway with soil. In the first jar place the plastic straw and glass marble. In the second jar, place the nail and aluminum foil. And in the third jar place the wood and fabric. Use masking tape across the front of each jar to list contents. Fill the jars with soil, completely covering the objects. Dampen the soil with a little water, and place the jars in a warm, sunny place. Do not cover the jars.

After 2 weeks, dig out the objects in the jars, spread them out carefully on the white poster board, and examine them carefully with the magnifying glass. Note any changes in the color, texture, smell, strength, and flexibility of each object. Which materials show signs of decay? Which appear unaffected? Record your observations carefully, then bury the objects again, and leave them in the moist soil for another week.

After the second week, dig them out. Do any of the materials that showed no sign of decay after the first week show any signs of decay now? Record your observations. It should be clear by now that some objects decay (biodegrade), and some do not. Among the objects that break down, some biodegrade more quickly than others. Begin to list objects in the order of their biodegrading properties, with materials showing the most pronounced decay at the top of the list.

Test your list's accuracy by burying the objects and leaving them in the moist soil for a third week. After that time, remove them and display them on a clean sheet of poster board for viewers, along with your observations and list. Have a magnifying glass handy for close examination.

Overpopulation Studies

You Will Need

- 32 guppies
- Fish food
- 3 small fish bowls
- Package of dry yeast
- Drinking glass
- Sugar-water solution
- Eyedropper
- Microscope

Overpopulation Effects

How many living things can our environment comfortably support? As the earth's population continues to grow, this question becomes increasingly important. A population out of balance with our planet's limited resources suggests future scarcity and starvation. It also means less privacy, reduced space, and massive waste. Overcrowding also affects how we feel about our fellow human beings.

To understand the effects of overcrowding, make a few casual observations. Study people on a crowded city street and make notes about their behavior. Ride in a standing-room-only bus or train and observe the postures, facial expressions, and body language of the people around you. Of course, these observations reveal only superficial and transitory effects of overpopulation since these people will soon leave these crowded environments. But for most people, lack of elbow room is indeed unpleasant.

Overpopulation in Guppies

Using human beings as subjects in an overpopulation study is impractical. Small animals, however, can clearly demonstrate the effects of overcrowding and provide valuable data and fascinating observations.

Fill each fish bowl three-fourths full of tap water. Allow the water to sit undisturbed for a few days before adding the guppies. Then, add 2 guppies to the first bowl, 10 to the second, and 20 to the third. Keep the bowls undisturbed and under the same light and temperature conditions. Add a pinch of guppy food to each bowl at the end of each day.

Study the behavior of the guppies in each bowl, and record your observations. Note the different breathing patterns—or gill movements—in the guppies from bowl to bowl. Why do the guppies in the crowded bowl spend more time near the surface and breathe more rapidly than the guppies in the less crowded bowls? Also compare guppy movement and aggressive behavior between bowls. Why do the guppies in the crowded bowl flick their tails less often and rarely swim except to chase other fish away? How do the fish behave at feeding time?

After a few weeks of observation, you should see a difference in the growth rate of the guppies, too. The bowl containing only two fish appears to have the largest and healthiest specimens, while the fish in the second and third bowls appear smaller and in poorer health. If you allow your experiment to proceed for a month or more, the guppies in the most crowded bowl will begin to sicken and die. After you've completed your observations, release the guppies into a pond or larger fish tank.

Overpopulation in a Yeast Colony

If you have a microscope, you can demonstrate the effect of overcrowding on microscopic organisms. Let's look at live yeast cells.

Buy a package of dry yeast and put ¼ teaspoon (1.25 ml) of it into half a glass of sugar-water solution. Wait about an hour, then use the eyedropper to place a drop of the solution on your microscope slide. Count, as accurately as you can, the number of yeast cells. Repeat this procedure every day for two weeks using the original solution, each time recording the number of yeast cells. You could demonstrate a **population curve** on a graph as the number of cells in the colony increases.

Notice what happens to the yeast population as its growth continues unchecked. Soon the population of cells outgrows the resources of the environment. Limited food, oxygen, and space, and the increasing toxicity of waste, doom the colony as more and more cells die off.

Light Pollution Diorama

You Will Need

For the Box

- 4 large pieces of foam board 40 inches (100 cm) square
- 1 large piece of navy blue poster board 30 inches (75 cm) square
- Duct tape (opaque)
- Transparent tape
- Tinted cellophane—red, yellow, and blue
- Flat black spray paint
- Black tempera paint
- Artist's paintbrush
- Household cement
- Modelling clay
- Small pieces of twig
- Artificial sphagnum moss

For the Lampposts

- 3 dowels 6 × ½ inches (15 × 1.25 cm)
- 3 pieces of ½-inch (1.25-cm) plywood 2½ inches (6.25 cm) square
- Three 6-volt flashlight bulbs
- 3 enamel flashlight bulb sockets
- Six 3-foot (90-cm) sections of insulated wire
- 6 alligator clips
- 4.5-volt battery
- 3 Ping-Pong balls
- ⅛-inch (.31-cm) plywood 2 inches (5 cm) square
- ⅛-inch (.31-cm) plywood 1½ × 2½ inches (3.75 × 6.25 cm)
- Spray paint can cover

Additional Materials

- Drill with ½-inch bit
- Coarse sandpaper
- Wire clippers
- Table vise
- Craft knife
- Carpenter's glue
- Small brads or nails
- Poultry lacing needles
- Constellation chart

Can't See for the Light

It might seem strange to think of light as a kind of pollutant. But just talk to the astronomers, who need the pure dark of night to make their best observations. As cities encroach upon once-rural settings, the lights from highways and parking lots, shopping centers and stadiums, even streets and backyards, have filled the sky with a glowing pinkish fog.

Some cities have taken steps to curb this problem by using yellow sodium lamps for outdoor lighting instead of the bright white mercury ones. The yellow of burning sodium gas is **monochromatic**; that is, the light has almost all the same wavelength and can be easily filtered out by astronomers.

Another approach has been in light design. Older lamp designs tended to favor little, if any shielding, and light was thrown in all directions. New designs attempt to fo-

Diorama Inside & Outside

cus and concentrate light where it's needed.

In this project, we'll build three lampposts and test each in a simulated starry night environment, the diorama.

Constructing the Diorama Box

Use a yardstick and a craft knife to cut the foam board into five panels (see illustration). Keep leftover foam board close by. We'll use it later.

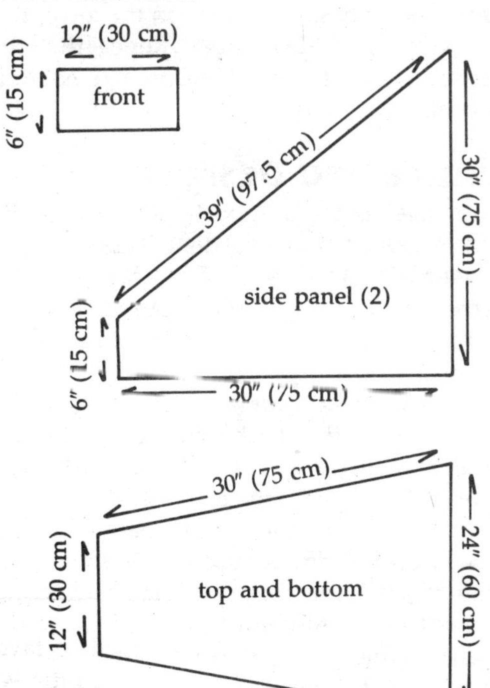

Take the rectangular front panel and determine the exact center by connecting the midpoints along the vertical and horizontal edges, and dividing the panel into four equal quadrants. The center point is where the quadrants converge. Measure 2½ inches (6.25 cm) on each side of the center point along the horizontal. From the ends of that line, draw 1½-inch (3.75-cm) vertical lines and connect them with a longer line. You should wind up with a rectangle 1½ × 5 inches (3.75 × 12.5 cm) centered on the front panel. Use the craft knife to carefully cut out the rectangle—the viewing hole of the diorama.

Assemble the box by first attaching the front panel to the bottom piece. Straight pins come in very handy for this. Use the pins to loosely connect pieces together and check the fit. When satisfied, apply a little fast-drying cement to the edges of the pieces and reattach them with the straight pins. Allow at least 20 minutes for the joint to dry before proceeding to the next step.

Attach the sides in the same way. When dry, use the flat spray paint on the interior of the box, blackening all four surfaces thoroughly. Then, paint one side of the top panel, and one side of any large pieces of leftover foam board. *Use spray paint in a well-ventilated area or outdoors. Make doubly sure that you carefully point the nozzle of the paint away from your face when you spray.* Ask an adult to help if you find spray painting difficult.

The navy blue poster board will bend to fit between the walls of the box at the wide end, but first prepare it with an attractive nighttime scene.

The Starry Night Sky

Paint a skyline along the bottom edge with black tempera paint. Use the silhouettes of rooftops, trees, chimneys, or telephone poles—whatever suits you. Avoid making your skyline more than 4 inches (10 cm) high, however. You want to have as much navy blue "sky" showing behind and above the lamppost as possible.

Next, carefully spread the poster board

on a flat surface with the skyline pointing up. After consulting the constellation chart for the position of constellations on and above the horizon (choose any season and compass point for the most conspicuous constellations), lightly dot the posterboard with yellow pencil. Or, you might challenge viewers to make real-life comparisons by observing the night sky at, say, 10 P.M. and re-creating a portion of it on your poster board. If you choose to do this, don't forget to record the positions of the planets.

Use a piece of surplus foam board for a pad as you begin punching holes in the poster board with a poultry lacing needle (for the constellations), and straight pin (for the auxiliary stars). Use the needle for planets as well, but tape a small piece of tinted cellophane behind each planet hole in the appropriate color—red for Mars, yellow for Jupiter and Saturn, and blue for Venus. Nonwater-soluble magic markers on a piece of clear tape work well, too. Finally, trim ¼ inch (.62 cm) from the top of the poster board using the craft knife and yardstick.

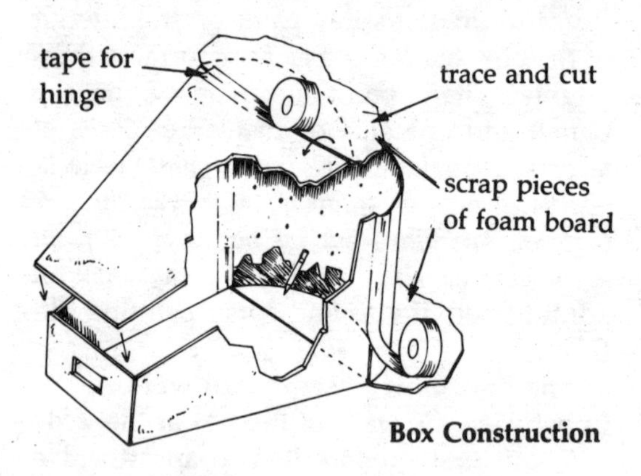

Box Construction

Assembling the Diorama

Carefully bend the poster board to fit between the walls of the box's wide side. It should billow out from the end of the box in a flattened semicircle. Tape the board to the walls both inside and outside the box. Then paint the inside tape black to match the walls.

Slide a large piece of extra foam board under the poster board and trace around the inside of the semicircle. Cut out the semicircle, glue around the edges, and push it against the bottom piece of foam board and poster board. Hold another large piece of foam board against the top of the poster board, this time tracing around the *outside* of the semicircle. Cut it out and attach it to the top panel with a long strip of tape, making sure the tape functions as a kind of "hinge" and the semicircle as a "flap."

Making Lampposts

We'll construct three model lampposts and test each one in the diorama for light pollution. All three share the same basic design, but with important differences.

Begin by drilling a ½-inch (1.25-cm) hole in the center of each 2½-inch (6.25-cm) square piece of plywood. Insert a 6-inch (15-cm) dowel rod into each of the holes, sanding the end of the dowel rod if necessary.

Use the table vise for the next part. Place a Ping-Pong ball in the vise and carefully tighten the grip. With the craft knife, carefully cut a ½-inch (1.25-cm) diameter hole in the ball. Repeat the procedure for the other two balls. Now, place a dowel in the vise so that the top sticks out about ½ inch (1.25 cm). Carefully saw a groove in the top of the dowel ¼ inch (.62 cm) deep. Repeat the procedure for just one of the two remaining dowels.

Unshielded Design

The ornamental lampposts of many public squares and commercial districts fall into the *unshielded* category. Globes, roofless lanterns, open grids, and exposed element lighting throws the light down, up, and all around.

To construct a basic unshielded lamppost, screw the flashlight bulb into the enamel socket, and carefully cement the bottom of the socket to the top of the first dowel post. If you have difficulty balancing the socket on the dowel, use a little modelling clay to keep the socket in place while the cement dries. Apply a small amount of cement around the edges of the hole you cut in one of the Ping-Pong balls, and gently drop the ball over the bulb, so that it rests on the socket. If you don't want the ball to permanently bond with the socket, substitute modelling clay for the cement, pressing a small string of it around where the ball and socket join.

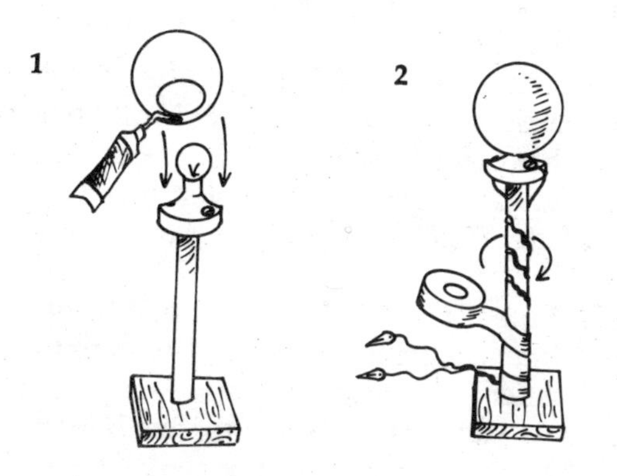

Unshielded Lamp Assembly

Attach two 3-foot (90-cm) wires to the terminals of the socket. Twist together the wires to about 1 foot (30 cm) from the terminals, and wrap the twisted wires, barber pole–style, around the dowel to the base. Give the wires an extra turn around the base before separating them. Finish the lamppost by wrapping the dowel with tape to hide the wires, attaching alligator clips to the ends of the wires. Paint your lamppost silver or grey for a more realistic effect.

Minimal Shielding Design

Most highway lights fall into this category, as well as your friendly neighborhood lamppost. A partial hood blocks out some light, but not much.

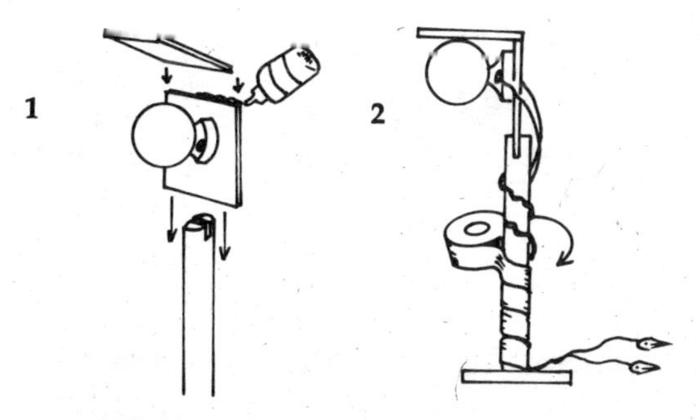

Minimal Shielding Lamp Assembly

To construct a *minimal shielding* model, attach the Ping-Pong ball to the enamel socket as before; then cement the socket to the 1½ × 2½-inch (3.75 × 6.25-cm) piece of ⅛-inch (.31-cm) plywood. Place it so that the smaller piece of plywood, or the hood of the lamp, rests perpendicularly on the ball for support. Join the two pieces of plywood together with carpenter's glue. When dry, insert this piece into the slot on the dowel, applying a little more glue, if necessary. Finally, connect wires, twist, and tape.

Full Cutoff Shielding

Once a rarity, full cutoff shielding has become more commonplace in parking lots, particularly around airports. This design directs bright light precisely where it's needed—*down*, with no light spilling to the sides or above.

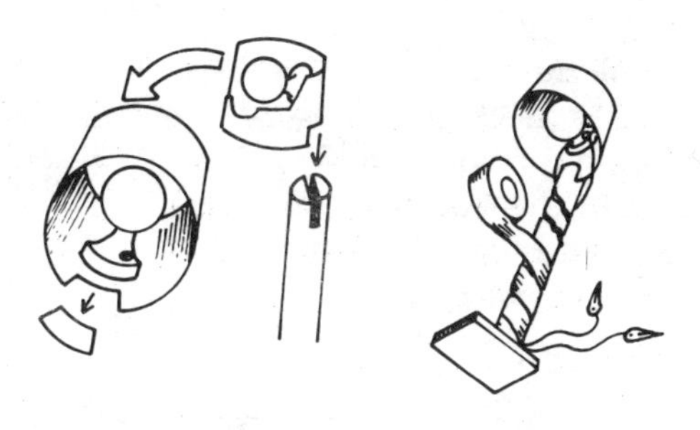

Full Cutoff Shielding Lamp Designs

For this model, clip a small notch on the edge of the spray paint can cover so that the cover can slip smoothly into the groove at the top of the dowel.

Attach a Ping-Pong ball to a socket and a socket to the inside of the can. You may have to build up the space behind the socket with a little cardboard before you can cement it in place. Slide the can onto the dowel, attach wires, and tape as before.

With the third of the set of lampposts completed, place the first lamppost inside the diorama about 4 inches (10 cm) from the poster board, centering it carefully. Make a small hole in the poster board at the bottom edge for passing wires through.

Trees & Bushes

Successful dioramas create scenes which aid in the communication of scientific fact. A natural history museum, for example, wouldn't have nearly the impact if prehistoric animals weren't displayed in a recreation of their natural environments. For this reason, spend some time landscaping the inside of your diorama, particularly around the lamppost. Use small twigs placed in modelling clay, and artificial sphagnum moss to create the effect of foliage. The moss comes in handy for covering the hole you made for passing through wires.

Pin and glue the top panel to the rest of the diorama. Attach a small strip of tape to the flap so that you can open and close it easily. Attach this strip around the flap with pins at first, then with more tape. Pleat the tape as you go around the curve to take up the slack.

Look through the viewing slit of your diorama to check for light leaks. The only light entering the box should come from the star holes in the poster board. If you see light coming through the seams, use a little modelling clay for repairs.

A First Peek

Position a strong lamp in *back* of the diorama so that light comes through the star holes. Attach the alligator clips of the wires to the terminals of a 4.5-volt battery, then look through the viewing hole.

You will see that the unshielded lamppost model clearly illuminates the ground and much of the bottom part of the diorama. But what can you say about the stars above it? How far above the lamppost must you look before the stars appear clear and brilliant?

Replace the unshielded lamppost with the minimal shielding model. How does the light differ in relation to star-viewing? Can you see stars lying closer to the horizon now? Can you make out the constellations clearly?

Finally, place the full cutoff shielding lamppost in the diorama. The effect is dramatic—a sky full of brilliant stars and pastel-hue planets above a well-lit ground. Can you see why this design is far superior to the others?

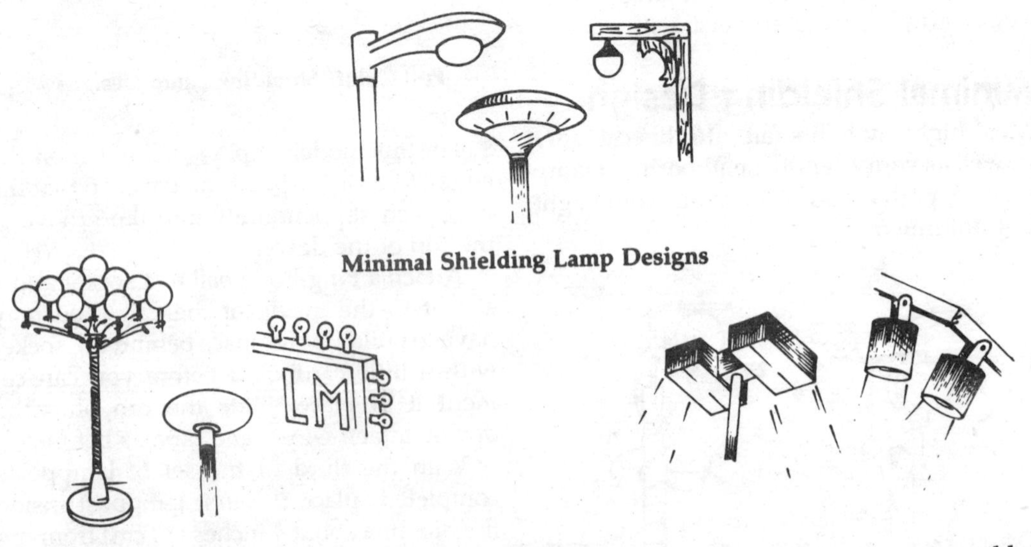

Minimal Shielding Lamp Designs

Unshielded Lamp Designs

Full Cutoff Shielding Lamp Assembly

Eye & Mind Tricks

Moving Cloud Forms
Mutoscope
Phenakistoscope
Zoetrope
Stroop Effect
Irradiation
3-D Stereoscopic Viewer

Moving Cloud Forms

You Will Need

- Phonograph
- White poster board
- String
- Hole punch
- Large picture of sky with clouds

Optical Illusions

Working together, the eye and brain provide information about our environment. But both eye and brain can be fooled, creating an **optical illusion**. Scientists study optical illusions for what they reveal about perception, and because some illusions can be dangerous. Pilots, navigators, and machine operators depend on visual information and must be alert when their eyes are likely to fool them. This project demonstrates how eye and brain interpret motion together and what happens when that perceptual mechanism becomes fatigued.

Spinning Spiral

Measure the phonograph turntable's diameter. Tie one end of the string a little more than half that diameter to a pencil. Using cellophane tape, attach the other end of the string to the middle of the poster board. Draw a circle by pulling the pencil tight against the string and swinging it around. Before you detach the string from the poster board, mark the center of the circle. Cut out the circle, punch a hole through the center mark, and place this paper "record" on the phonograph turntable.

Switch the record player to 45 rpm or medium speed. As the record spins, hold a felt-tip marker close to the center and slowly drag it out to the edge. You will create a perfectly drawn spiral. To draw a spiral in the opposite direction, flip the record over. This time, begin at the edge and slowly move the marker towards the center.

Place the turntable in front of a wall about 6 feet (1.8 m) away. Tape the cloud picture to the wall, and clearly illuminate it. If possible, increase the phonograph speed to 78 rpm. Look down at the turntable and stare

Cloud Poster & Turntable

at the spinning spiral for about 30 seconds. For 45 rpm, stare at the spiral about 1 minute.

Now quickly look up and stare at the cloud picture. The clouds appear to move. This effect persists about 20 seconds even if you look away and back again. Notice, too, that the clouds seem to move in a direction opposite that of the spiral. If the spiral moves inward, the clouds appear to drift towards you. If you flip the record over and reverse the spiral's direction, the clouds appear to move away.

What's Happening?

Receptors in the eyes work with the brain to detect *inward* and *outward* motion. When you look at stationary objects, inward and outward receptors are in balance. But when you look at a spiralling pattern, the stimulation makes one set of receptors tired. As you stare at the cloud picture, *resting receptors* take over, and that's why you see motion in the opposite direction.

This works best with objects with vague or complex outlines. That's why cloud forms demonstrate this effect dramatically. But trees and rock formations work almost as well, and you could try substituting other pictures.

You could use a lazy Susan instead of a phonograph turntable. With the lazy Susan you can just reverse the direction of spin for the second spiral. However, this requires someone to spin while you stare at the spiral, and the spinner must maintain a steady rate of rotation.

Mutoscope

You Will Need

- Small stiff notepad
- Crayons or marker

Making Movies

Before the invention of the motion picture camera, many odd devices brought still pictures to life, creating the illusion of motion. One such device, the **zoetrope**, meant "wheel of life." Other devices were the **mutoscope,** or flip book, **phenakistoscope** (which means "deceiver"), and **kinetoscope,** or picture strip. Thomas Edison invented the kinetoscope for use in nickelodeons—the direct forerunners of the motion picture projector.

The principle for all these devices is the same. Separate, sequential images flash by at high speeds, fooling the brain into seeing one continuous image in motion. Each separate image persists in the brain about 1/16th second. Images flashing by at speeds greater than 1/16th second blend into a single moving picture.

Mutoscope

We'll construct the simplest device, the mutoscope. Draw a simple design on every page of a small, stiff notepad, starting at the back. Make each design relate to the one before it. If you begin with a square and want to turn it into a circle, for instance, gradually round off the square's corners in the next few drawings. If you want your square to revolve, tilt the square a little more in each drawing, allowing your drawings to flow naturally one into another. This ensures a smooth image when you flip.

When you finish your drawings, flip through the pad like a deck of cards, from back to front. Your design comes to life. Construct several flip books with more complex images.

Phenakistoscope

You Will Need

- Masonite board 1 foot (30 cm) square
- Wooden dowel ¼ × 8 inches (.62 × 20 cm)
- 2 wooden balls or ornaments to fit dowel's ends
- Drill with ¼-inch bit and 3⁄32-inch bit
- Old 10-inch (25-cm) record
- 2 wire coat hangers
- 1-foot (30-cm) square plywood for base
- 1-foot (30-cm) square freestanding vanity mirror
- Knife file
- Wire clippers

Sophisticated Deceiver

Slightly more sophisticated in design than the mutoscope, the **phenakistoscope** provides livelier images. The design also allows more flexibility in the kinds of drawings you can do.

Wheel Construction

Trace an outline of the record onto Masonite board. Have a lumberyard cut the 10-inch (25-cm) circle from Masonite and smooth the edges. Using a ruler, divide the circle into twelve equal pie slices. With a saw, cut 1-inch (2.5-cm) slots from the circle's edge, following the pie slice lines. Where lines converge at the circle's center, drill a ¼-inch (.62-cm) hole. Widen the 1-inch (2.5-cm) slots with the knife file. Make each slot no more than ⅛ inch (.31 cm) wide, just comfortable enough to peek through.

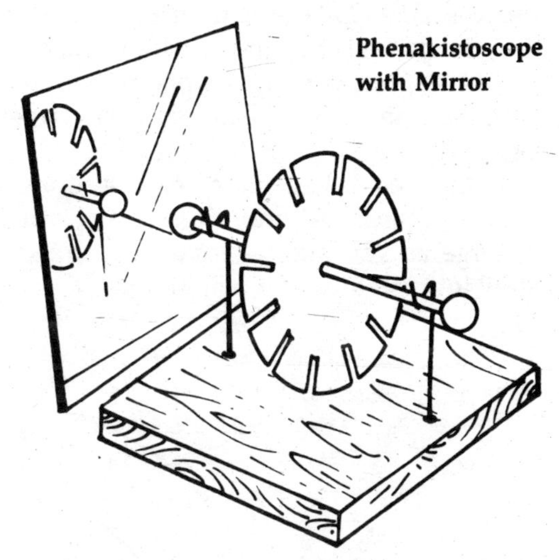

Phenakistoscope with Mirror

Push the dowel through the center hole until it pokes halfway through the circle. Make sure you have a snug fit. Attach a wooden ball to one end of the dowel.

Base Construction

With wire clippers, snip two 1-foot (30-cm) lengths of wire from coat hangers. Measure 1½ inches (3.75 cm) from one end of each length, and use pliers to bend the wire into a hook-like shape. Remove any excess wire with the clippers. Using a pencil, divide the plywood base into two halves. Along this center line, drill two 3⁄32-inch (.24-cm) holes, separating them by 6½ inches (16.5 cm). Insert the straight ends of wire into these holes, and twist the wires until their hooks are parallel. Suspend the wheel and dowel mechanism from the hooks and spin it. It should rotate smoothly.

Drawings

Use the circular diagram as a guide for your phenakistoscope drawings. Cut several 8-inch (20-cm) diameter circles from paper. Use a pencil compass to measure the circles or trace around a record's edge. Draw three more circles inside the main circle—one with a 5½-inch (13.75-cm) diameter, another with a 3½-inch (8.75-cm) diameter, and the smallest with a 1-inch (2.5-cm) diameter.

The diagram includes a series of X's off-center in circular configurations. The outer ring contains 12 X's—the number of slots on the wheel—and the inner circles contain 13 and 11 X's, respectively. Duplicate this configuration in light pencil on each paper circle.

In the area designated by each X, draw a design or small picture. Make the drawing as large as you like, as long as it doesn't stray into the area of a neighboring X.

When you've completed your drawings, poke a hole in the circle's center, and push it along the dowel. With tape, attach the circle to the Masonite wheel. Place the second wooden ball on the dowel's end to keep the wheel from slipping off the hooks.

Making It Work

Place your finished phenakistoscope on a small table with the paper circle facing a vanity mirror. Look towards the mirror through a slit on the Masonite wheel. Spin the wheel and enjoy.

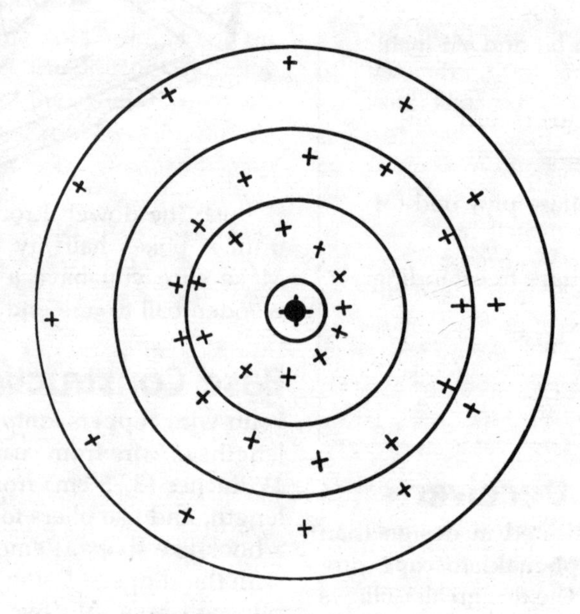

Pattern for Phenkistoscope

Zoetrope

You Will Need

- Clear plastic cylindrical cake cover
- 5-inch (12.5-cm) diameter ball-bearing turntable
- Wood block 5 × 5 × 5 inches (12.5 × 12.5 × 12.5 cm)
- Plain 3 × 5-inch (7.5 × 12.5-cm) index cards
- Plywood for platform 8 inches (20 cm) square
- Transparent acetate sheet
- Plexiglass glue
- 4 small wood screws ¼ inch (.62 cm) long or less
- 4 machine bolts with nuts

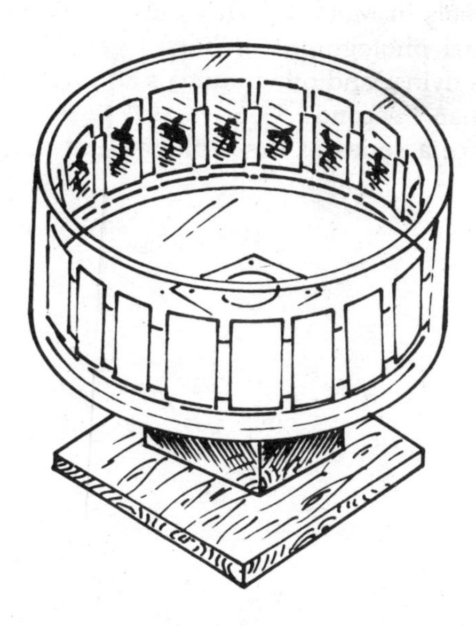

Zoetrope with Stand

Parlor Amusement

In the 19th century, the zoetrope was a popular parlor amusement. When used with sequential photographs, it produces clear, smooth images that remind us of early motion pictures.

Preparing the Cards

Begin with about 15 index cards, depending on the cake cover's circumference. Turn them narrow side up, and number each card inconspicuously in the corner.

Carousel and Card Holders

Turn the cake cover upside down and straddle it between two books so that it sits level. Use small pieces of tape to attach the index cards around the inside of the cover until they're evenly spaced and separated by no more than 1/16 inch (.16 cm). You may have to trim the last card. With a waterproof marker, draw a line along the top and bottom edges of each card. Then, remove all cards.

To make card holders, cut the acetate sheet into 1 × ½-inch (2.5 × 1.25-cm) strips, twice as many strips as cards. Carefully apply plexiglass glue to each strip's long edge before pressing them against the waterproof-marker lines, with unglued edges facing each other. Allow the glue to dry.

Turn the cover right side up, and carefully remove the handle using a screwdriver or small hacksaw. Sand the surface to remove bumps and scratches. Center the ball-bearing platform on the cake cover, and mark the screwhole positions with a grease pencil. Drill over the marks and attach one side of the platform to the cover with machine bolts and nuts. Attach the other side of the platform to the wood block using wood screws.

Be sure that you attach the swivel to the exact center of the cover. If it is off-center by even a fraction of an inch, this will create problems with balance when you spin the carousel.

Tracing Old Photographs

Early photographers took detailed photographs of animals and people in motion. A famous collection by Edward Muybridge is "Animals in Motion." Muybridge took sequential photographs of horses galloping, doves flying, and other animals at intervals less than .1 second.

Trace a series of these photos and glue your traced drawings onto index cards. Since you've numbered the index cards, this will help you attach the drawings in proper sequence. Carefully slide each card between the two acetate strips until you completely cover the inside of the carousel. The rest is easy. Spin the carousel while looking through the side.

Stroop Effect

You Will Need

- 17 plain white index cards
- White poster board
- Felt-tip markers—red, yellow, blue, orange, green, purple
- Stopwatch

Perceptual Paradox

When the eye and brain are confronted with conflicting information, we have a **perceptual paradox**. The Stroop effect examines this paradox. The test involves several people reading color words written on incongruent colors—the word *RED* written in *yellow* letters, for example. Results are timed and compared for word and color recognition.

You'll need at least 20 subjects—10 males and 10 females—and someone to test and time them.

Preparing the Cards

Place the stack of index cards in front of you. Use 6 different felt-tip markers to write a color word on each card according to the chart shown. Leave one card blank to cover the stack when you begin the test.

Word	Color of Marker
red	yellow
yellow	blue
blue	red
orange	green
green	purple
purple	orange
red	green
yellow	purple
blue	orange
yellow	red
blue	yellow
red	blue
green	orange
purple	green
orange	purple
green	red

The first test involves color recognition. In preparing the test, choose an area where other subjects cannot view the current test subjects.

Place the stack of cards on a table with the subject in front of it. When the subject feels comfortable, remove the cover and start the stopwatch. When the subject names the color of the top card, pull that card away to reveal the card beneath. Continue until the subject identifies the color of the last card. Stop the stopwatch and record the total time.

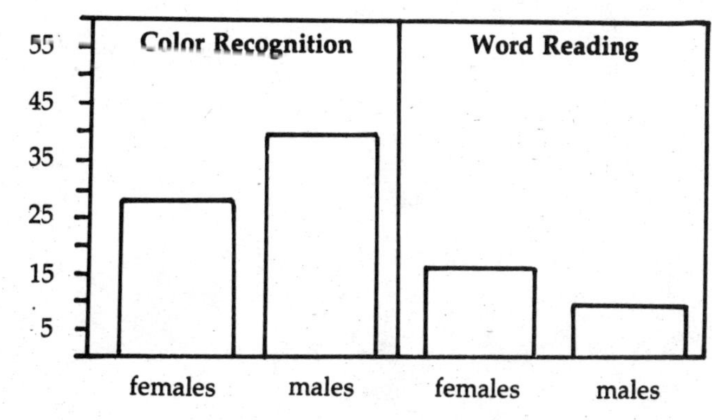

If the first subject is female, repeat the test with the remaining females. Then average results by adding all their times together and dividing by the number of females tested. Repeat the test using males; again, average results and record them. Finally, repeat the test following the same procedure, but have subjects read the color *word*. Average results as before.

Surprising Conclusions

Comparing the average times of males and females, you'll discover that females recognize colors slightly faster than male subjects. Males, however, recognize words slightly faster than females. And both males and females recognize words much faster than they do colors. Draw a chart to plot results.

When confronted with a perceptual paradox, the brain processes symbolic information more quickly than sense information. Other studies have suggested that females have a slight advantage in color recognition, and males are slightly better at symbol recognition. Whether those differences are innate or learned is uncertain.

Irradiation

You Will Need

- White poster board
- Black paint
- String

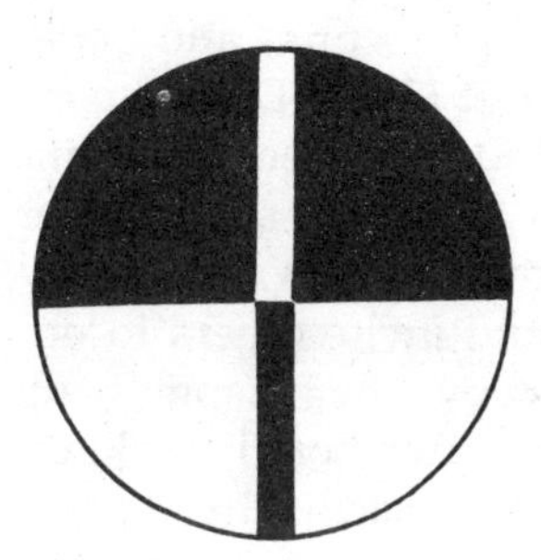

Retina

The **retina**, the inner part of the eye, is covered with light-sensitive cells. Some of these cells, **rods,** distinguish light from dark. When contrasting shades of dark and light are placed very close together, the eye becomes confused, particularly where contrasting shades meet. This is the spillover, or **irradiation** effect.

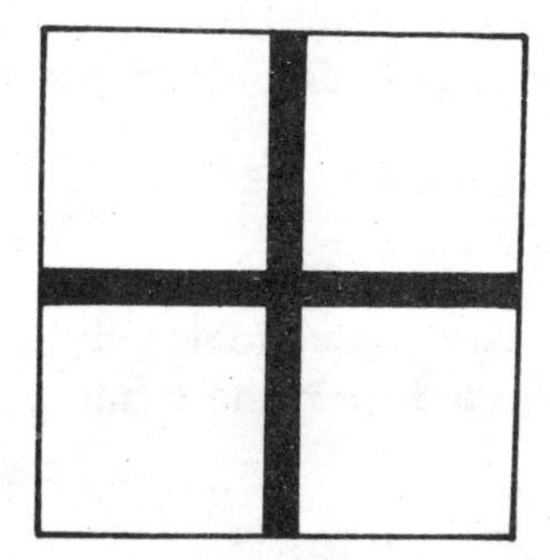

Op Art

A similar effect exists between certain colors, like blue and orange. When they're placed directly against each other, the outline of one color shape against the other appears indistinct and unstable. Blue and orange excite such different types of color-sensitive cells, **cones**, that when they are placed together the eye has difficulty shutting off one set of cells and switching on another. Putting a black line or strip of white between the colors allows the eye to relax and adjust.

That's why certain color combinations are avoided for road signs, billboards, and other things that must be easily read. But this effect can create interesting art and advertising images.

Making Pictures

We'll demonstrate black-and-white irradiation with pictures. Copy the circle shown on a piece of poster board. To draw it accurately, tie one end of a string to a pencil and tape the other end in the middle of the

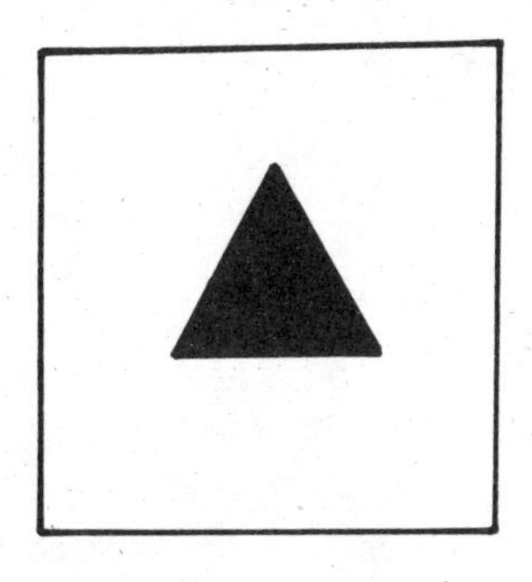

Irradiation Patterns

poster board. Pull the pencil tight against the string and swing it around in a circle. Use a ruler to divide the circle into equal halves, and make sure the vertical center strip is the same width, top to bottom.

Hang your picture, and stare at it from about 10 feet (3 m). Does the circle appear evenly divided between black and white areas? Does the center strip appear to have a uniform thickness? Write down your observations, and invite others to do the same.

Now copy the triangles on separate pieces of poster board, making sure both triangles are exactly the same size. Hang your pictures, step back, and record your observations. Finally, copy the cross and hang it beside your other pictures. How does what you see differ from the other two images?

Fooling the Eye

In every case, you've fooled the eye. Light waves reflected from the white part of the image excite those parts of the retina on which the image is formed and neighboring parts as well. The brain "sees" more white than dark, although the areas are exactly the same.

In the first picture, the white stripe appears much wider and longer than the black stripe. But at the same time, the white half of the circle appears larger than the black half. So, you "see" the white stripe extend down into the area of the black stripe, crossing the horizontal line of the circle.

The second picture shows how different two identical triangles look when painted in contrasting shades. The white triangle looks about 15 percent larger than the black one.

Finally, if you stare at the third picture, you'll see the black lines retract from the paper's sides, an effect similar to that observed in the first picture. Design eye-fooling pictures of your own.

3-D Stereoscopic Viewer

You Will Need

- 2 small pocket mirrors, same size
- 2 pieces of thick plywood, cut slightly smaller than mirrors
- Thick plywood for base
- 4 rectangular wooden blocks
- 2 strips of thinner wood or Masonite, about 5 × 12 inches (12.5 × 30 cm)
- 4 metal L-shape brackets and screws to fit
- 10 wood screws 1 inch (2.5-cm) long
- Epoxy glue

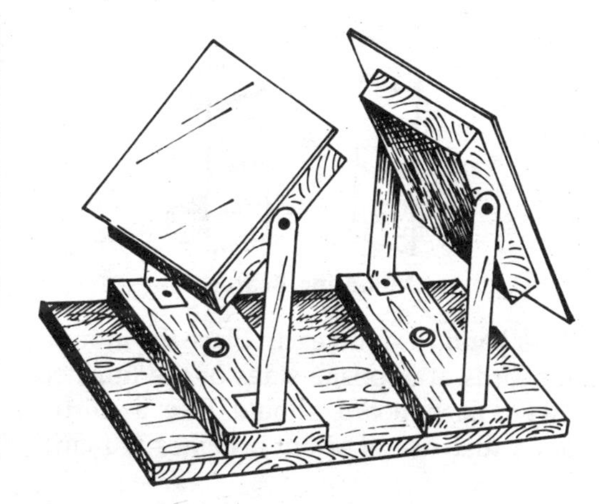

3-D Stereoscopic Viewer

Depth Perception

The best drawing or photograph does not reproduce the world as we see it. The essential difference is **depth perception**. We can thank the combined efforts of both eyes for that. The brain receives two slightly different images from each eye and fuses them together to make one image. The result allows us to perceive our surroundings in *perspective*.

Using the same principle, this stereoscopic viewer, or **stereoscope**, simulates depth perception using homemade drawings.

Making the Stereoscope

To construct the mirror units of your stereoscope, center and glue the plywood pieces to the backs of the mirrors so that no more than ¼ inch (.62 cm) of mirror protrudes along each edge. Allow several hours for the glue to dry. Measure the longer sides of each plywood piece, and make a pencil mark along the edge at the midpoint. Screw the long end of each L-shape bracket to the pencil marks with the shorter end facing in. Then screw the shorter ends to the wooden blocks. Make sure the brackets line up so that the mirrors are level.

Measure the long sides of the plywood base, and divide the surface in half with a line. On each side of that line and parallel to it, draw another line, ¼ inch (.62 cm) away. Place the mirrors upside down, top edges against these outside lines, so that ½ inch (1.25 cm) of space separates them. Mark the plywood base—both on the surface and along the edges—with X's where the brackets connect to the sides of the mirrors.

Flip the mirror units over, placing the wooden blocks between the X's. Make sure the mirrors are no more than ½ inch (1.25 cm) apart when tilted perpendicular to their brackets.

Measure the length and width of each wood block, and mark the exact center. Attach the blocks to the base with 1-inch (2.5-cm) wood screws through the marks, making sure each block pivots freely from side to side. Both mirrors should also pivot between their brackets. This is important for making viewing adjustments later on.

To make two mounting boards for the stereoscopic pictures, screw long strips of thin plywood or Masonite against the sides of the remaining wooden blocks. Stand the boards upright on their blocks, and loop two rubber bands around each. Place the mounting boards at opposite sides of the finished viewer. Each board should face a mirror from a distance of about 8 inches (20 cm).

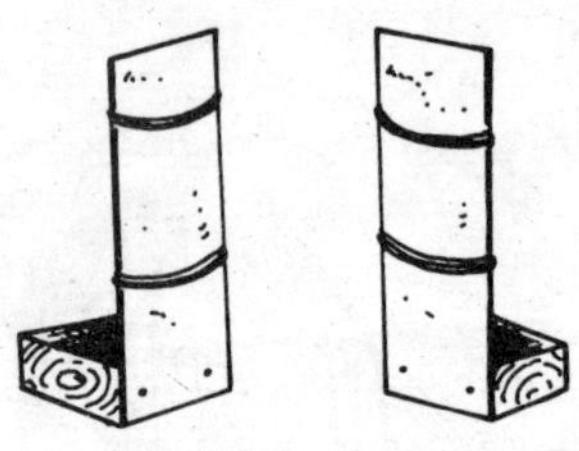

Mounting Boards

Creating a 3-D Image

When you view the two drawings shown through a stereoscope, they combine to form a single three-dimensional image. Notice the four planes of perspective in each drawing—fence, horse, barn, and clouds—and how the horse, barn, and clouds have been moved to the right in drawing B.

These drawings make up a **stereoscopic pair**. You can easily make your own pairs from a single drawing. In the drawings shown, you could substitute anything for the objects—low bushes, cows, distant trees, airplanes. But you must have *foreground, middle distance, background,* and *distant background* in your first drawing.

Look again at the farmscape. Drawing B was traced from drawing A in stages. The fence was traced exactly in position. Then, tracing paper was moved ⅛ inch (.31 cm) to the left, and the horse was traced. The barn was traced after moving the tracing paper another ⅛ inch (.31 cm) to the left, and finally a fourth plane was added by moving the paper another ⅛ inch (.31 cm) to the left and tracing the clouds.

Try colored pencils for your drawings. Color enhances perspective and often suggests its own perspective. Blue, for instance, *appears* to be behind red or yellow on a flat surface.

Stereoscopic Photos

You can create another stereoscopic pair with two photos. For a close-up, still life, or portrait, take the first photo with the subject about 2 feet (60 cm) from the camera. Move the camera 3 feet (90 cm) to the right (without turning it) and take a second photo. The first photograph represents the left eye's view and the second, the right eye's.

For more distant subjects, estimate the distance between the camera and the nearest object, say, a house, and move your camera to the right about one-eighth of that distance for the second photo. If the house sits 100 feet (30 m) away, move the camera 12½ feet (3.75 m) to the right for the next photo.

A

B

Pair of Stereoscopic Drawings

Stereoscopic Vision Test

Attach your drawings (or photos) to the mounting boards to test the viewer. Place your original drawing on the right board, securing it with rubber bands. Place your traced drawing on the left board, securing it the same way. Position both drawings at the middle of the mounting board and brightly illuminate them with desk lamps.

Tilt the viewer's mirrors away from each other at a 45-degree angle. Sit in front of the viewer and bring your head down over it, with one eye over each mirror, until you see two separate drawings completely filling the mirrors. Now carefully adjust the mirrors (remember, the mirrors pivot against their brackets as well as against the base) until the drawings overlap and the space between the mirrors disappears. The effect is miraculous.

You could draw the first half of many stereoscopic pairs and invite viewers to create the second half by tracing the first. Make sure you have enough tracing paper and colored pencils to go around.

Index